The University of Chicago Press,
Chicago 60637
© 2022 by the Trustees of the Natural
History Museum, London
Published 2022
Printed in China

31 30 29 28 27 26 25 24 23 22   1 2 3 4 5

ISBN-13: 978-0-226-82430-7 (cloth)
ISBN-13: 978-0-226-82431-4 (e-book)
DOI: https://doi.org/10.7208/chicago/9780226824314.001.0001

First published in the United Kingdom in 2022 by the Natural
History Museum, Cromwell Road, London SW7 5BD.

Library of Congress Control Number: 2022934877

Contents

Foreword

Almost every conversation with Sandy Knapp opens with the deliciously enticing phrase "did you know…" followed by an inexorable outpouring of fascinating erudition. These wonderful essays brim over with all the delight of a wandering Sandy conversation: witty, intriguing, educational and inspiring. Coupled with a series of stunning images, Sandy's book is both a pleasure to browse and a joy to read.

There is, however, a serious point here too about taxonomy or the science of classifying organisms. Only by collecting, conserving and classifying the marvellous variety of life on Earth can we begin to understand how life evolves, how it responds to changes in our natural environment and, sometimes, how and why life goes extinct and disappears. Taxonomy is the science which underpins all the crucial scientific research into life on Earth. It is, if you like, the vocabulary and grammar of our description of the natural world.

So taxonomy matters, and scientists have been seeking to describe and classify the huge variety of life on Earth for at least the past 2,000 years. Different approaches have been used over the years but all modern classification systems have their roots in the Linnaean classification system developed by the Swedish botanist Carl Linneaus in the 1700s. Linnaeus's system classifies organisms into an ascending hierarchy from species to genus to family, order, class, phylum and finally kingdom. But perhaps his greatest contribution was his method of giving each species a two-word Latin name consisting of the genus name and the species name, with just these two words sufficient to uniquely identify every species on Earth: hence *Homo sapiens* or 'wise human', although these days some may dispute the moniker.

Sandy's book takes the brilliant concept of interweaving plants, people and names by focusing on those people who the scientists of the day regarded so highly as to use their names to identify, not just a single species, but an entire genus. Here you will have the opportunity to meet some very particular humans and their stories. Some well known like Queen Victoria or Benjamin Franklin, others perhaps more surprising like Sequoyah or Lady Gaga! All fascinating stories and beautifully brought to life and linked to the plant genera in Sandy's elegant prose.

Finally of course, no book about natural history can ignore the dire threats to the state of the world's biodiversity. If Sandy's book can help to inspire a little more love and care for the natural world then her efforts will not have been in vain.

Doug Gurr, *Director, Natural History Museum London, 2022*

Banksia formosa from Western Australia used to be classified in the genus *Dryandra*, but taxonomic work showed it was nested within the banksias (see *Banksia*) – the previous genus name lives on in its common name, the showy dryandra.

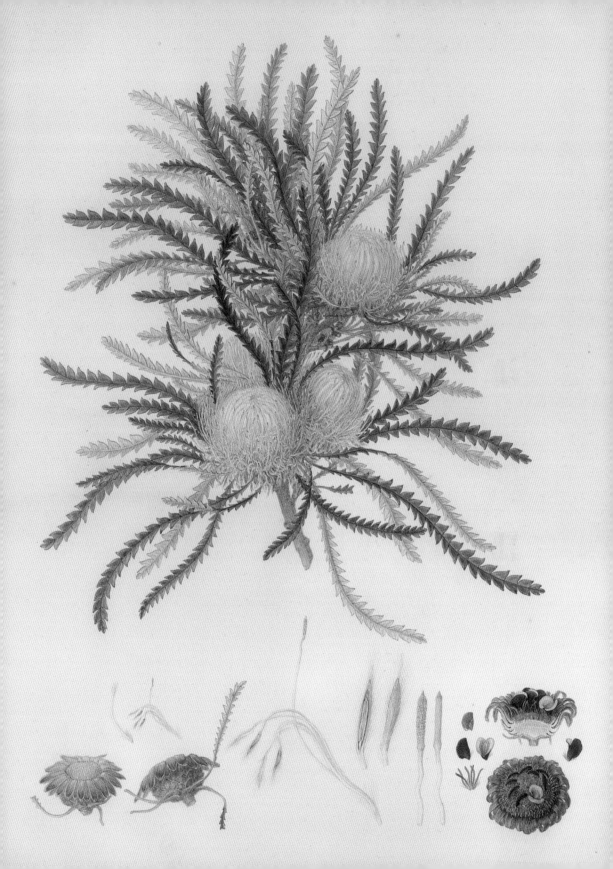

Introduction

Plants and people have a long association with one another. Plants are essential for human survival, they generate the air that we breathe, they form the basis for ecosystems we exploit, and they provide us with food, fuel, fibre and forage. Ever since people began to use plants, they also named them; names are essential for communication, and the usefulness of plants meant they needed to be talked about. Different cultures across the world each developed naming systems for the plants around them, often focusing on the organisms' properties or uses. All this worked well until humans began to travel further and further from home, exploring, colonizing and conquering new lands. In large part driven by European colonization, unknown plants from very far away began to be shared. The local names of plants arriving in Europe from China or Brazil, say, became lost in transit, and Europeans took up the process of naming each plant anew. This was the genesis of the scientific naming system we use today.

By the late 17th and early 18th centuries Greek and Latin were the languages of European scholars, so it was logical that when Europeans began to coin names for new plants, they did so in these langauges. This created a universal language of names that transcended local naming traditions, allowing all botanists to communicate with one another. At its inception, this system was limited to Europeans – short shrift was paid to the traditions of other cultures, except where it might benefit the colonizers. So, plants were named in these ancient languages for their characteristics or their uses, and, as more diversity was encountered with increasing exploration and exploitation, for people as well.

I have always enjoyed the scientific names of plants; as a taxonomist, I am immersed in them, coining them, sorting them out, interpreting them, saying them and feeling the Latin syllables roll off my tongue. There are stories in plant names, about the plants themselves, where they grow and sometimes who they are named for. Sometimes the meaning is clear – *Solanum arboreum* means the 'tree-solanum' – the species name *arboreum* tells us the plant is a tree. The name is a representation of the plant, but on its own cannot tell the entire story. This is especially true when the plant is named for a person. On its own the name *Magnolia* does not tell us of Pierre Magnol's life or of the influence he had on botanists at the time he lived, or even that the plant was named for him at all. I am fascinated by the intersection of stories that comes from plants named for people – this book is the result. I have focused on the

The antimalarial drug quinine comes from the bark of *Cinchona*, a member of the coffee family named for the Countess of Chinchon who, legend has it, was cured by its use in 17th century Peru.

Cinchona officinalis. L.

IN THE NAME OF PLANTS

names of plant genera, in part because there are fewer of them than there are species, but also because naming a genus is perceived as a bigger step than naming a species, perhaps necessitating more thought or consideration. It is also true that the genus name comes first; you can't name a species without it having a genus to belong to!

By convention, the scientific naming system we use today for plants officially began on the 1st May 1753, with the publication of the Swedish botanist Carl Linnaeus's *Species Plantarum*. It's not that names had not been coined before this, just that botanists needed to pick a point in time after which they would all follow a set of rules which they had mutually agreed. This has become the *International Code of Nomenclature for algae, fungi and plants*, most recently agreed in Shenzhen, China, in 2017; it will next be revised in Madrid in 2024.

Concepts of the genus and species were circulating widely in the communities of early European botany – a genus was considered a kind of 'basket' into which one of more species were placed, the genus was a broader, more inclusive category. Each species was distinguished by a phrase describing it. As more plants came to be known, these phrase names, as they were called, became longer and longer, making them difficult to keep track of. In *Species Plantarum* Linnaeus codified the use of a one-word name for the species, resulting in binomial nomenclature – each species had a name composed of two words, a genus name and a species name, or epithet. The system was easy, designed as it was as a memory aid, and it obviously worked well since it is still in use today. Thus, we have names like *Magnolia grandiflora* and *Magnolia stellata*, much easier to remember than long sentences describing these two species.

But before Linnaeus published *Species Plantarum* he already had very clear ideas about the right and wrong ways in which to name plants. Never one to wait on another's opinion, in 1736 he published a book outlining his fundamental ideas of how to study plants, called of course *Fundamenta Botanica*. He expanded this a year later with *Critica Botanica*, in which he laid out the reasoning behind his rules and, twenty years later, refined these rules – what he called 'aphorisms' - in *Philosophica Botanica*. Here, along with how to name plants, he described plant development, structure, and basically everything you wanted to know and more.

Many of the aphorisms about generic names relate to naming plants for people. Only certain types of people merited this honour in Linnaeus's eyes. So, for him, genus names were not meant to be thrown around willy-nilly, but reserved for gods, poetic subjects, and the luminaries and promoters of botany. Kings were all right since they had the potential to give botanists money! Linnaeus kept many of the names that had been coined by earlier botanists, like those of the Frenchmen Charles Plumier and Joseph Pitton de Tournefort, but rejected names coined by others like the London apothecary James Petiver, who, he said "thrust priceless gifts, too brilliant

The blue pincushion *Brunonia* was named in honour of Robert Brown who collected it in Tasmania in January 1804, where its beauty was recorded by the artist Ferdinand Bauer – today we would take a photograph!

Robert Brown (1773–1858) accompanied HMS *Investigator* to Australia as the ship's naturalist, and later became the first Keeper of Botany at the Natural History Museum, London.

and valuable for the uneducated, on florists, monks, relations, friends and the like. Gifts thus given are no gifts, but only a source of derision and scandal to posterity." In his *Critica Botanica* he enumerates lists of generic names honouring people, making some dubious connections that had not been made by the originators of those names. He justifies the naming of plants for botanists by stating that in other professions names are "customarily attached to their accomplishment" – so why should botanists labour in vain? "Under what unlucky star was our unrewarded science born?" A bit melodramatically, he goes on to list botanists who sacrificed much for their science (including himself, of course).

A name is, however, only a way to communicate about something. And a genus is to some extent a matter of opinion – if we think of the tree of life as a branching diagram, and genera as one level of branching, where that level occurs is based on evidence a scientist has available at the time. One botanist might put several species into a single genus, while another might name them as different genera, all from the same branching diagram. Today we have many more sources of evidence at our disposal when we make the decisions what to call any given plant; each name is a hypothesis, capable of being falsified by new evidence. This is what makes taxonomy a science and not just a bookkeeping exercise.

For this book I have selected plant genera that are named for different kinds of people – there are so many genera named for people that some readers will surely be offended at my omissions. Many more plant genera are named for men than for women or for indigenous people for example, and I am very pleased that my questions to colleagues about plants to include in this book have sparked an effort to verify all plant generic names that honour women, and link these to or even create their biographies on Wikipedia. I have tried to include some genera that will be obvious to readers, like *Victoria*, and others that may have new stories to tell, like *Gaga*. In each portrait I have also tried to convey some of the extraordinary beauty and variety of the plant world, and how much we still have left to learn, even about things we think we know a lot about. Plants and people both have their stories. I have also included genera that are being named by botanists today, because of course naming plants is not just an activity that happened in the far distant past. I hope that this book will open windows into the plants and the people for whom they are named; sometimes they are intimately linked, as Linnaeus would have wished, other times the link is more tenuous. We are still finding out so much about the world, all the time.

I found out so much writing this book – about plants, people, and I had some exciting moments too. I went to Traveller's Rest in the mountains of Montana, USA where Meriwether Lewis and William Clark collected a specimen of *Lewisia*, I found the 'lost' type specimen of *Darwinia* in the herbarium, I got to explore the Museum's extraordinary art collection. But most of all, in writing each chapter

Helia Bravo Hollis (1901–2001) was the world specialist on the cacti of her native Mexico; the genus *Heliobravoa* was named in her honour, as were two Mexican botanical gardens.

I learned so that much I never knew about these amazing plants and the people for whom they are named, both the person and the plant came alive for me through their interconnected stories.

Sadly, plant diversity is under threat, not just in far-away places but in our very backyards. The twin drivers of climate and land-use change are irreversibly altering our world, and the plants upon which we so depend. As Sir David Attenborough (for whom the genus *Sirdavidia* is named) eloquently said in a radio interview in the lead-up to the 2021 Climate Conference of the Parties (COP26): "The world is dependent on plants, yet we treat them with so little respect or care." Naming something gives it space, power and legitimacy; both scientific and local names tell stories when you delve into them. Names can serve to connect us to plants in a very special way, these connections are even more important now that so many plants are in danger of being lost forever.

Sandra Knapp, *London, 2022*

Adansonia

MICHEL ADANSON

Families: Malvaceae, the mallow family
Number of species: 8
Distribution: Africa, Madagascar and Australia

One of the goals of taxonomy and systematics is classification – organising the lineages of nature we study into groups that correspond to what we consider 'natural', groups. Today we recognize that organisms belong to lineages that share characters (features) inherited from their antecedents; it is telling that the only illustration in Charles Darwin's paradigm-changing book *On the Origin of Species by Means of Natural Selection* is a diagram of ancestry and descent, a tree of life. The 20th century German fly taxonomist Willi Hennig formulated the study of definition and recognition of these lineages into a method he called cladistics, where lineages were defined by their possession, not just of characteristics, but of shared, derived characteristics.

In the latter part of the 20th century, new methods for arriving at the tree of life – identifying those lineages whose numbers were all each other's closest relatives – were hotly debated in scientific circles. Cladists, who were using the methods first articulated by Hennig, were challenged by pheneticists, who insisted that purely numerical methods would be the best way to elucidate relationships. They insisted that all characters be created equally – shared possession from a common ancestor was not necessary. In these debates passions ran high. In heated controversies it is often the case that scientists set up historical figures as heroic forerunners of their ideas, suggesting that the ideas they advocate are really the modern development of something from the past. Those advocating purely numerical methods (the so-called pheneticists) identified the French botanist Michel Adanson as the precursor of their methods – calling their approach to classification 'Adansonian'.

Adanson was a student of Bernard de Jussieu, botanist at the Jardin des Plantes in Paris in the mid-18th century and contemporary of the great Swedish botanist Carl Linnaeus and, like much of the world of botany at the time, immersed in bringing order to the diversity of plants people were finding beyond European shores. Linnaeus had devised a system for classification that relied on counting the numbers of male and female parts in the flower; this worked well for placing an unknown plant with others like it, but the system was highly artificial. In order to find a natural system,

The five-parted leaves of *Adansonia digitata* give the species name to this plant – these leaves, botanically called palmate, look a bit like a hand with its five digits.

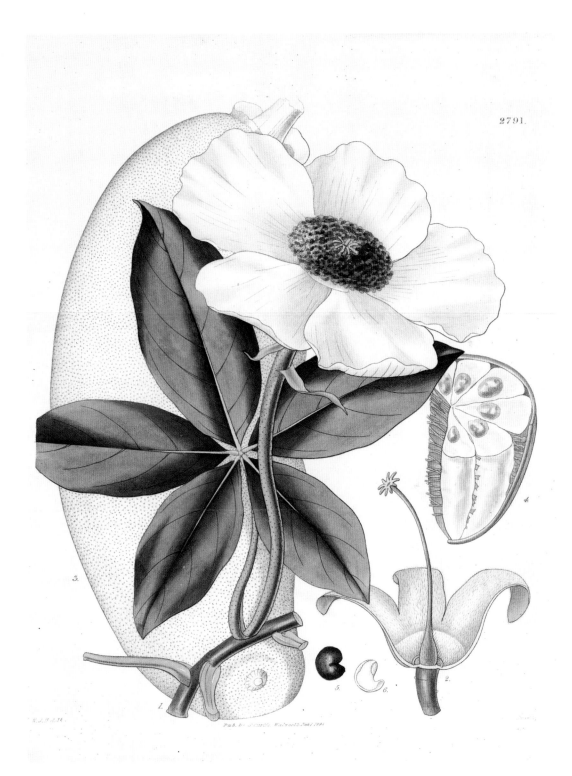

2791.

botanists began exploring other ways to classify. Debate raged between different schools of thought. In 1763 Adanson published his *Familles de Plantes*, in which he looked at many features of plants to uncover their affinities. His view of a natural classification was that it should take into account all features of organisms, not just a few. It was this use of many features that the pheneticists latched onto; it sounded just like them! But it wasn't really. Adanson used many features, certainly, but he also stated that some characters were better than others and that characters varied even within his natural families. His was not a purely mechanical, numerical method. He went beyond what many of his contemporaries had done and, rather than just listing the component genera in families or orders, he described them, with their characters and how they varied. He distanced himself from the idea that groups had essences and could be identified with single features. Some of his groupings sound familiar today – 'Les Orchides, *Orchis*' (Orchidaceae), 'Les Composees, *Compositae*' (Compositae) and 'Les Malves, *Malvae*' (Malvaceae).

It was in this last grouping that he placed the genus Linnaeus had named for him, adding "Le Baobab est vraisembleablement le plus gros de végétaux" [The baobab is apparently the largest of plants]. Adanson was not a purely theoretical botanist, working with systems of classification and searching for ways to organise knowledge about nature; as a young man he had travelled to western Africa, sent in the field because he was too young at 20 years old to be admitted to the Académie des Sciences, despite his precocious interest in and talent for all things scientific. Funded by the director of the Compagnie des Indes (a French trading company set up the century before to compete with the English and Dutch), Pierre-Barthelemy David, he set off for Senegal, then under French domination, not as a botanist in the pay of the king or as an official naturalist, but as a clerk. Senegal at the time was relatively unknown to Europeans; save for the infamous Gorée Island that had, in 1677, been taken from the English by the French. The island, off the coast opposite today's capital Dakar, was the centre of the trade in humans undertaken by all the European powers from the 15th to 19th centuries. Today it is a UNESCO World Heritage Site, serving as a reminder of human exploitation, and is a sanctuary for reconciliation. In Adanson's day, it was contested territory – the trade in human lives was big business. He hid many of his maps and descriptions of the interior of the country from the British so that they

MICHEL ADANSON
(Botaniste)
Membre de l'Académie des Sciences,
Né à Aix (B.-du-Rhône) le 7 Avril 1727,
Mort à Paris le 5 Aoust 1806.

When he returned from Senegal in 1754 Michel Adanson (1727–1806) tried to publish his observations, but the book made huge losses, leaving him in penury for the rest of his life.

did not regain possession of the territory. Adanson spent five years, from 1749 to 1754, in Senegal, collecting, documenting, and observing. He not only collected widely but grew plants in his garden to test the characteristics given to them by Linnaeus, and to compare with plants already known. He didn't confine himself to plants either, he collected mammals, birds, insects, fish (pressing them flat between pieces of paper as suggested by the botanist Konrad Gesner), but also languages, customs, and mores of local people.

As a young man who still felt he had much to learn, he was reticent about corresponding with the great botanists of the day, like Linnaeus, and did so through Bernard de Jussieu, his mentor and teacher. Jussieu did just that with the extraordinary tree Adanson found – the baobab. Adanson wrote to Jussieu "I will leave it to you to communicate to Linnaeus the character of the Baobab; I don't think this will be disadvantageous to me or have any bad consequences; and I would like you to assure this learned man the great esteem in which I hold him and his works". As perhaps a precaution for his young protégée Jussieu sent the plant to Linnaeus with the name 'Adansonia digitata' – cementing Adanson's role in its discovery for science.

It is not clear where Adanson first saw baobab trees – he appears to have been on a hunting trip when he first saw what we now know as *Adansonia digitata*:

> ".. I laid aside all thoughts of sport as soon as I perceived a tree of a prodigious thickness, which drew my whole attention. [...] There was nothing extraordinary in its height; for it was only about 50 feet; but its trunk was of a prodigious thickness. I extended my arms, as wide as I possibly could, thirteen times before I embraced its circumference; and for a more exact measure I measured it round after with a thread and found it to be sixty-five feet; and therefore, the diameter was about twenty-two. I do not believe the like was seen anywhere in the world; … "

He was the first European botanist to describe the baobab in its native habitat; the tree had long been known to Europeans for its large fruit, whose dried pulp and seeds were highly prized medicinally. Local people in Senegal and throughout Africa were well aware of this tree; its importance stretched beyond use as a food and medicine but into the realm of religion and mysticism, as it still does throughout the range of the genus.

Although *Adansonia* was first described botanically from the species on the African continent, the great species diversity is found in Madagascar. All species of *Adansonia* are what we call pachycaul trees – in reference to their thick trunks, so remarked upon by Adanson. There is some evidence that the diameter fluctuates with rainfall, so it may be that water is stored in the trunk. The branches of baobab trees are short and stubby looking compared to the trunks, giving

The pulp surrounding the seeds in the huge fruits of the baobab (this one is 35 cm/10 in long) is highly prized for its sweetness and nutritive value.

This patch of *Adansonia rubricaula* in Madagascar looks for all the world like the trees have been pulled up and popped back in upside down.

the trees a slightly comical, incongruous look. In Australia the one species that grows there, *Adansonia gregorii*, is sometimes called the upside-down tree; when the leaves fall during the dry season, the branches certainly look a bit like roots. The huge size of these trees is also reflective of their great age; these are among the oldest of flowering plants. One tree in Zimbabwe was radiocarbon dated to 2,450 years old when it died in 2011, and several others in southern Africa are estimated to be more than 2,000 years old.

One of the iconic images of Madagascar is the grand avenue of *Adansonia grandidieri*; these trees have become almost a symbol for the extraordinary flora of that island, where so much of the diversity occurs only there. But neither great age nor iconic status can protect these wonderful trees from environmental changes wrought by humans. In Madagascar the great avenue of baobabs is at risk from water-logging due to encroaching rice cultivation and, in continental Africa, baobab die-off is thought to be the result of climate change. As with much of biodiversity, these extraordinary trees face an uncertain future, as do the rest of the organisms that depend upon them.

In this tree of superlatives, the flowers live up to the rest of the plant. Baobab flowers are large, fleshy, and open only at night; flowers buds can be up to 29 cm (11 in) long and the open flowers, with their central brush of stamens, are as big as your hand when open. The flowers of the African species, *Adansonia digitata,* is the only baobab whose flowers dangle down below the branches on long stalks – perfect bat-flowers. In the early 20th century pollination of tropical trees by bats was not thought important, but observations of baobabs in botanical gardens showed that this was possible and in fact common. Fruit bats are the pollinators of *Adansonia digitata*; they visit the sour-smelling creamy-white flowers that open at dusk to obtain the copious nectar held at the base of flower tube. In so doing, they brush against the stamens and stigma, picking up and transferring pollen as they fly from tree to tree.

Previous classifications of *Adansonia* grouped all the species with the stamens borne at the ends of long narrow tubes together; this meant the Australian and four of the eight Malagasy species were thought to be closely related. But, as Adanson himself said, single characters often vary in unpredictable ways. Reconstruction of the evolutionary history of *Adansonia* using DNA sequences showed that this was not so; all of the Malagasy species, those with and without long staminal tubes, were most closely related to each other. The similarity was due to retention of a character related to hawkmoth pollination, while the shorter tubes had evolved independently in the African baobab and in two of the Malagasy species (*Adansonia grandidieri* and *Adansonia suarezensis*). Amazing fieldwork by a botanist perched in baobab trees showed these two species in Madagascar were pollinated by lemurs and fruit bats respectively. The floral features of these species with upright flowers and those dangly flowers of the African species are due to convergent evolution and are not shared derived characters. This does not mean they are useless to us as botanists, they tell us a lot about the ways these extraordinary trees live their lives.

In Antoine St. Exupery's iconic small novel *Le Petit Prince* (*The Little Prince*) baobabs stand for all that is evil on the prince's planet; grown too large they can break the planet itself –"It is a question of discipline ... When you've finished your own toilet in the morning, then it is time to attend to the toilet of your planet, just so, with the greatest care. You must see to it that you pull up regularly all the baobabs at the very first moment they can be distinguished from rosebushes that they so resemble in their youth." The prince is right, one must attend to the health of the planet, but not by uprooting the iconic baobab trees. They have been with us for millennia and have seen much, they deserve our care and attention into the future.

The long tube formed by the stamen bases in the flowers of the baobab means the nectar held within is only available to animals with long tongues who can reach it.

Agnesia

AGNES CHASE

Family: Poaceae, the grass family
Number of species: 7–8
Distribution: South America, except the Andes

Convincing a non-botanist that grasses actually have flowers can be a challenge. Agnes Chase, lauded by her peers as "the dean of American agrostology" (the study of grasses) attributed her love of the grass family to a childhood incident in which she brought a bouquet of grasses to show their tiny flowers to her grandmother who insisted that grasses did not have flowers. She later recalled the incident saying "I was right and she was wrong."

And she was right, grasses are flowering plants and have extraordinary flowers, pared down to the minimum, tiny but perfect. Look closely at a spike of wheat or the tiny fluffy inflorescence of the grass growing in pavement cracks and you will see what Agnes Chase saw. Grasses have inflorescences, groups of flowers, often composed of many hundreds of individual flowers – called florets in grasses. The florets themselves are held in spikelets – any plant family with such specialized and different-looking structures usually has a terminology all its own. But bear with me…. each spikelet has, at its base, two bracts that are called glumes, then each floret has two bract-like structures that look a bit like opening jaws called the lemma (the lowermost one) and the palea (the upper one). These two structures often have conspicuous awns – bristly tips that can be very long and obvious. Within the palea and lemma is the flower itself – simplicity personified, just two or three stamens with anthers that dangle in the wind and a feathery stigma that protrudes to catch the windblown pollen. At the base of the stamens and ovary are three tiny bump-like projections; these lodicules are thought to be homologues of petals in other flowers. So truly a flower like no other. Of course, there are variations on this basic theme in grasses – sometimes there are bristles within the flower, sometimes the terminal or last florets are sterile, with no reproductive parts. All of these differences help in the identification of grasses, so notoriously difficult for beginners. Agnes Chase realized this, and her beginners guide to grasses, first published in 1921, was designed to open the world of grasses to everyone. As well as being the world's expert on grasses and grass taxonomy, she wanted to bring her knowledge and expertise to as many people as possible – today we would say she was active in science communication.

Mary Agnes Chase was one of five children brought up by a widowed mother, first in rural Illinois, later in Chicago, where the family moved after her father died when

Agnes Chase collected this herbarium specimen of *Agnesia ciliatifolia* near the Brazilian town of Dourado in Matto Grosso in 1930. Her notes – "weedy border of matta" – tell later botanists that it grows in scrubby forest.

she was two years old. She went to school, but then needed to help make ends meet. At nineteen she worked as a proof-reader and typesetter for a mildly-socialist magazine, the *School Herald,* dedicated to helping rural schoolteachers, where she met and subsequently married the editor, William Ingraham Chase – an idealist who wanted to better the world. He was older than she, at thirty-four nearly twice her age, and was already extremely ill with tuberculosis and died less than a year after they were married.

To clear his debts and make a living for herself, she took on more proof-reading jobs, this time at a Chicago newspaper. As time went on, she proof-read for money, and botanized in northern Illinois and Indiana for fun, in her spare time. While out botanizing she met the Reverend Ellsworth Jerome Hill, who was studying local mosses. He discovered that Agnes had a flair for illustration, so persuaded her to illustrate his papers in exchange for lessons in botany and microscope use – much needed for the study of both mosses and grasses! Through Hill she met the botanist Charles Millspaugh of the Field Museum of Natural History in Chicago, who had her illustrate (for free, mind you) two of the Museum's publications on the plants of Central America.

Clearly with her advancement in mind and in light of her considerable skill as an illustrator, Hill convinced Agnes in 1903 to apply for a position as botanical illustrator at the US Department of Agriculture (USDA) in Washington D.C. She was ranked first of all the applicants and, although somewhat reluctant to have applied, took up the post with alacrity; it started her on her illustrious career as a grass specialist. She illustrated publications for the Forage Division, probably mostly grasses, and after-hours – again in her spare time – she began a study of the panic grasses. She started to work with Albert Spear Hitchcock – the USDA agrostologist, first as an illustrator, but soon as a valued colleague and equal. By 1907 she was appointed as scientific assistant in systematic agrostology. She was certainly not a subordinate however; her letters to Hitchcock from the field always began "Dear Prof. Hitchcock" but were full of observations on grasses, instructions on what to do with things back at the herbarium and general comment on the world. They read as letters between extremely good friends. A colleague remarked "It is doubtful if Prof. Hitchcock could have accomplished as much as he did except for the constant interest and assistance given by Mrs. Chase."

Although much of her work was in the grass herbarium of the USDA, Chase was not a "closet botanist", as her colleague Jason R. Swallen put it in his tribute to her on her 90th birthday in 1959. She went on field trips all over the United States, spending 1908, 1910 and 1912 in the western states, gaining first-hand knowledge of the grasses in the field, important for the agricultural and rangeland mission of her institution. In 1913 she made her first 'foreign' collecting trip – to Puerto Rico; the *Grasses of the West Indies*, co-authored in 1917 by Hitchcock and Chase, relied heavily on her field observations. When Hitchcock requested that some of his stipend for exploring the newly acquired Panama Canal Zone be given to Chase to fund her own field work, a Smithsonian bureaucrat showed just what women scientists were up against – "I doubt the advisability of engaging the services of a woman for the purpose [of exploration]."

How wrong he was. Agnes Chase was an extraordinary field botanist – nowhere more obvious than in her two extended collecting trips to Brazil, the first in 1924-1925 and the second five years later from 1929 to 1930. Brazil had been long ignored by American grass taxonomists, who relied on old collections held in European institutions. Chase set out to redress the balance. In the course of her two trips, she collected more than 4,500 specimens of grasses plus many other plants besides; she added something like 10% to the world's knowledge of Brazilian grasses with her collections – "I shall have made, in these two trips, by far the largest collection of grasses ever brought out of Brasil." Her sense of fun and good humour shine through in her letters to Hitchcock from the field – along with a genuine desire to assist botanists in Brazil. In a letter sent during her second trip she lamented the state of literature availability for her Brazilian counterpart in the USDA.

Her letters from the field are peppered with social comment as well as botanical observations – on meeting a Russian emigrant who had fled the Bolsheviks and argued strongly for monarchy she

At home in both the field and the herbarium, Agnes Chase (1869–1963) worked in the collections at the Smithsonian every day for twenty years after her official retirement.

acerbically noted, with her tongue firmly in her cheek, that perhaps the emigrant would be better off in the United States where people knew their place and were imprisoned for protesting the status quo. And she would know about that – Agnes Chase was a committed suffragist, twice imprisoned and subsequently force-fed for public protests at the White House and House of Representatives in Washington D.C. She showed, through her actions, that women were as capable as men of strenuous field work – a photograph from her 1924 trip shows her and the Brazilian botanist Maria Bandeira atop the Agulhas Negras (the black needles) of the Itatiaia range near Rio de Janeiro in floppy hats and skirts over their trousers. This, after scaling the needles on all fours and at the end with ropes. In November 1929 she ascended the Pico de Bandeira with the collector Ynes Mexia:

> "…it was terribly steep. The bamboo (Chusquea trinii) tripped and caught us and showered us with water, though it had stopped raining – and hope deferred maketh the heart sick – it was up and up and up stumbling and crawling for an eternity. My knees got so wobbly. … but at last at 3:30 we struggled out of the bamboo and saw the men resting on the campo. I shouted for joy and old Antonio grinned and said something about "muito courageous" for senhoras to make that ascent. He said no woman had ever done it before and only a very few men."

It is fitting therefore, that a Brazilian grass genus bears Agnes Chase's name. *Agnesia* was described only in 1993, 30 years after her death. The genus was based on a species previously classified in the large and heterogenous bamboo genus *Olyra* and distinguished from it by fine structures of the spikelets and florets. New genera are often described this way, through recognition that part of a larger genus warrants

One might be forgiven for thinking this plant, *Agnesia ciliatifolia,* is not a grass at all. These small bamboos are often misidentified as dayflowers or gingers.

distinction because it is so different. More recently, *Agnesia* has been expanded to include more species previously thought to belong to *Olyra*, including one widespread species, *Agnesia ciliatifolia*, that Agnes Chase herself collected some 30 times in various parts of Brazil.

Agnesia is one of the 'herbaceous bamboos' of the Amazonian rainforests; we usually think of bamboos as large and treelike, but these little bamboos are smallish herbs of the dark rainforest understory and don't even look much like grasses – I have mistaken them for ferns. The understory of tropical rainforests is an unusual habitat for a grass. While the canopy or edge of a forest can be windy, the understory is usually rather still, with little air movement. So how do these understory grasses manage to get pollinated? Most grasses are wind-pollinated and rely on the breeze to carry the pollen grains from one plant to another, hence those dangling stamens and feathery styles. Studies on other herbaceous bamboos – but not yet on *Agnesia* – have shown that insects are frequent visitors to these plants. Flies, stingless bees, and beetles are attracted to the tightly packed florets with their bright yellow anthers and carry pollen on their bodies. The larvae of hoverflies (Syrphidae), are usually predatory, but a few tropical species have larvae that feed on the pollen of these herbaceous bamboos; in laying eggs females carry pollen from one plant to another. There is much still to learn about tropical grasses – something Agnes Chase would have loved.

That 'human grass' who devoted her lifetime to the understanding of grasses was also a stalwart supporter of other botanists, especially those from the American tropics. Her home was open to them, and many students spent extended periods of time studying grasses in the United States National Herbarium at the Smithsonian, where Agnes Chase worked every day, including Saturdays. She accomplished much even in the years after she retired. In 1951 she completely revised Hitchcock's *Manual of Grasses of the United States*; known as 'Hitchcock and Chase' it is the bible for agrostologists. She also completed a 'database' of all grass names on over 80,000 index cards. This was published in time for her to see it before she died at the age of 94, five months after she stopped working and on her first day in a nursing home. Her love of grasses was more than a list, it shone through everything she did; when in despair at lack of good habitat in the field she wrote "My spirits went up when I saw a stretch of clean grass country."

Banksia

JOSEPH BANKS

Family: Proteaceae, the sugarbush family
Number of species: *c.* 170 (incl. *Dryandra*)
Distribution: Australia, with one in New Guinea

Every Australian child has heard of the 'big bad Banksia men' the baddies of May Gibbs's early 20th century *Snugglepot and Cuddlepie* series, in which the characters are all based on native Australian plant species, usually those found in Western Australia where Gibbs grew up and played as a child. The Banksia men are based on the appearance of old cones of banksia species, with their open follicles as mouths and eyes, and the persistent flowers as the hairy bodies. Recalling a walk in Western Australia with her cousins, "We came to a grove of Banksia trees and sitting on almost every branch were these ugly, little wicked men that I discovered, and that's how the Banksia men were thought of."

The open gaping mouths of the Banksia men are the open seed pods of individual flowers; banksias have tightly packed inflorescences of many flowers, and the fruiting structures are referred to as cones. The follicles of each flower often open after fire, scattering seeds on the fertile, ash-enriched ground in the fire-adapted ecosystems where banksias grow. Most children probably do not know they are saying a scientific name when they read the stories of the Banksia men and their nefarious deeds, but the genus *Banksia* was named for Joseph Banks, whose relationship with Australia was both good and bad; almost as bad as those big, bad Banksia men.

Joseph Banks was an Englishman born to a life of privilege. His family were landowners in the fens of Lincolnshire, but Banks was born in their London home in Soho in 1743. Banks loved pottering outside and was not a bookish boy – only when the study of natural history became possible did his mind click into action. His status as a member of the landed gentry allowed him the freedom to pursue his natural history hobbies – both at Eton and Oxford he spent more

The gaping mouths on *Banksia* cones are the dry fruits of individual flowers, called follicles. Not every flower develops into a fruit with seed, but it only takes one to replace the parent plant.

On his return in 1771 from his voyage with Captain James Cook aboard HMS *Endeavour*, Joseph Banks (1743–1820) was full of ideas but never got around to publishing the results of his botanical labours.

time botanising than studying the classics. His independent, entrepreneurial streak was evidenced early on as he was allowed to accompany voyages of discovery – at his own expense of course, but with the moral backing of the Royal Society. He first went to Newfoundland in 1766, and famously later accompanied Captain James Cook on the voyage of HMS *Endeavour* to observe the transit of Venus from Tahiti and thence to the south to ascertain the shape and existence of the southern continent, collecting natural history specimens along the way. Early maps had it as a large land mass the size of Eurasia, almost 'balancing out' the land masses in the southern and northern hemispheres. Was it really 'Terra Incognita'?

Banks took along on the *Endeavour* not only a botanical companion Daniel Solander, who had been a pupil of the great Swedish botanist Linneaus, but also the draughtsman Herman Spöring, two artists Sydney Parkinson and Alexander Buchan, four servants and the best scientific collecting gear available - no small feat to fit all that on a relatively small ship along with the full complement of sailors needed to manage the ship itself, and to last the whole three year voyage still all on speaking terms. After successfully recording the transit of Venus in Tahiti, Cook then took the *Endeavour* south, where Banks met his eponymous plant genus for the first time in what is now called Botany Bay in late April 1770. There he and Solander collected three species of what would be called *Banksia*, all of which were painted by the talented Sydney Parkinson. Banks's party was so large and unusual, and his presence so commanding that once they were back in England the voyage was referred to as Mr. Banks's voyage, rather than Cook's. Banks's natural history collections were the talk of the town and parties came to the London house in Soho Square to goggle at the artefacts and specimens from far-off lands. The *Endeavour* voyage made Banks's name – he was elected to the Royal Society, became the darling of London society and in general had his opinion sought on everything. His fingers were in every pie.

He unsuccessfully attempted to join a second voyage to Australia; his grandiose scheme to enlarge the ship to accommodate his party caused the ship to keel over. Off Banks went in a huff to other things, like a possible trip to Russia, or a voyage to Iceland. But his life after the *Endeavour* really revolved around London and its clubs, societies and young ladies. He had the ear of the king, George III, who sought Banks's advice on all things botanical, including the establishment of his botanical gardens at Kew near

Richmond. Banks also had considerable influence in the decision to send the First Fleet to colonize Australia. The American War of Independence (or Revolution, depending on whose side you were on!) of 1776 meant that the freeing up of space in English prisons by transporting criminals to the east coast, of what was becoming the United States of America, was no longer an option for the British government. Another solution to prison overcrowding had to be found. The harshness of penalties for crime in the 18th century meant that many, what we might call petty, crimes committed by the poor were punishable by hanging but were often transmuted to transportation. New South Wales, that land eulogized by Banks and Cook, was seen as a place where the problem of prison overcrowding could be solved and turned to the benefit of Britain economically. Banks's testimony to a Parliamentary Committee was enough to set the wheels in motion. Once transported though, the people who went to Australia found no land just waiting to be turned into a replica of the English countryside, but instead hard poor soil, little water and little more freedom to prosper than the life they had left behind.

Banks, by now President of the Royal Society, was the champion of colonization of Botany Bay, even while the French Revolution occupied the government's attention in the latter part of the 18th century. He financed the transport of "Trees, usefull plants and Seeds" to the new colonies and began to send plant hunters to further explore the botanical novelties of Australia. Among these was the botanist Robert Brown who, with Ferdinand Bauer on the HMS *Investigator* captained by Matthew Flinders, explored the southern coast of Australia finding more of the amazing species of *Banksia*, which by that time had been described by the son of Carl Linnaeus (also called Carl) in Banks's honour: "In honour of Joseph Banks, the future author of the splendid work of all the plants detected in the lands of the southern seas with descriptions and figures." Said splendid work was not published until the 1980s, when all of the 743 plates prepared from Parkinson's illustrations made on the *Endeavour* voyage were finally published. The copper plates of Parkinson's paintings and sketches made on the voyage were engraved by a team of artists hired by Banks upon his return and, although the planned *Florilegium* publishing the discoveries of his botanising over the three years of voyaging was a dream for Banks, his concentration wavered onto his many other interests and endeavours and the project was not even close to being finished in his lifetime.

Banskia is, with the exception of a single species in New Guinea, endemic to the continent of Australia. It, along with the eucalypts, is an iconic part of Australian vegetation and ranges in form from mat-like shrubs creeping along the ground to trees 25 m (82 ft) tall. All species have complex, showy flower heads composed of many tiny flowers with narrow petals and copious nectar, which is a food source for many animals, humans included – most banksias are pollinated by either birds or mammals who rely on the sugar for fuel. But some have outrageously interesting pollination systems – like *Banksia epimicta*, whose foul-smelling flowers are subtended by dull brown hairy bracts that are thought to resemble a dead bird lying on the ground, just the thing to attract its blowfly pollinators! Banksias are adapted to fire; survival in the fire-prone habitats of Australia is either be by re-sprouting from a woody base – the 'sprouters' – or by fire-induced opening of the follicles, releasing thousands of seeds onto fertile ground – the 'seeders'. The inspiration of the Banksia men is thought to be the Western Australian 'seeder' *Banksia aemula*. Although banksias are adapted for fire, if it is too often or too hot, fire can kill both the sprouters and the seeders.

So imagine the consternation when a team of scientists in Australia decided that, based on the evidence from DNA sequences, the concept of that iconic entity *Banksia* needed to be enlarged to include the plants previously known as *Dryandra* (a genus named by Robert Brown for a colleague of Daniel Solander's and another pupil of Linneaus, Jonas Dryander). This flew in the face of the traditional classification of the two groups as distinct genera, which were considered closely related but easy to distinguish. So what was the problem then? At issue was the way in which taxonomists define taxa, such as genera. The goal of classification is the identification of what we call monophyletic groups – all the descendants of a common ancestor. The problem with *Banksia* and *Dryandra* was that the lineage defined as *Dryandra* was indeed monophyletic, but it was a branch from within *Banksia*, making *Banksia* what we call paraphyletic – a group that does not contain all the descendants of that common ancestor; without including *Dryandra*, *Banksia* was not a monophyletic group. Another way of saying this is that *Dryandra* was nested within *Banksia*. This situation occurs with the birds and dinosaurs – birds are a monophyletic group, but dinosaurs, without the inclusion of birds, are not. Both birds and dryandras have undergone a lot of morphological change over time, adapting to their environment in special ways. This can mean they look quite different to their close relatives, despite what the data tell us about their relationships.

One of the main differences that those wishing to retain the traditional two genera focused on was inflorescence shape – banksias have elongate, taller flower groups that look more like bottlebrushes, while the dryandras have flattened shorter inflorescences with a ring of enlarged leafy structures at the bottom, making them look a bit like chunky daisies. But there are species traditionally called *Banksia* that have short cones, like the dryandras. The traditionalists also feel that not enough species were included in the DNA analyses, nor was the whole genome used. But the argument really rests on the acceptance or non-acceptance of paraphyly in classifications. Name changes upset apple carts but are often a sign that we have learned something new about the group in question. So what to do about *Banksia*? There are usually several solutions to such a problem – one is to recognise a more inclusive *Banksia* that comprises all of the descendants oꜰ the common ancestor, while another is to recognise many smaller groups, each with a new name. Australian botanists have chosen to do the former and

This painting of *Banksia spinulosa*, the hairpin banksia, was done by an unknown artist known as the 'Port Jackson Painter' soon after the transport of the first European colonists to Australia.

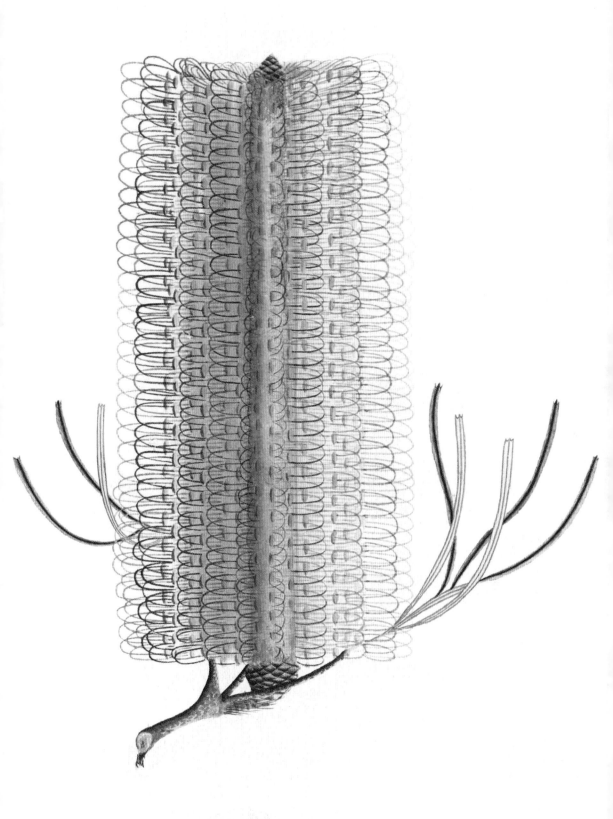

The flowers of the Western Australian bird's-nest banksia, *Banksia baxteri*, begin to open from the bottom of the large cone-like inflorescence that provides a long-lasting nectar reward for the plant's honeyeater pollinators.

now the official plant lists for the country accept the inclusive definition of *Banksia*. This meant many changes in names, for there were many more dryandras than banksias. Those changes happened more than a decade ago now, despite the continued complaints of traditionalists that "there is no obligation to follow the change simply because it is the latest word, or because herbaria have adopted it." True, but it helps us to communicate if we use the same names, and if the scientific evidence of evolutionary relationships tells us a story that is at odds with the traditional view, isn't it time to re-examine the evidence?

Joseph Banks died in 1820, knowing that one of the extraordinary plants he and Daniel Solander had collected in Botany Bay was a plant genus new to science that bore his name. He had bequeathed his herbarium of dried plants and his library to the British Museum (part of which is now the Natural History Museum, London) via his then librarian Robert Brown; he was remembered by his colleague Sir Humphry Davy as "a good-humoured and liberal man, free and various in conversational power, a tolerable botanist and generally acquainted with natural history. ... He was always ready to promote the objects of men of science, but he required to be regarded as a patron and readily swallowed gross flattery." I think Banks, the more than "tolerable" botanist, would have preferred the large inclusive *Banksia* that encompasses much diversity and evolutionary novelty to the words of his erstwhile colleague.

Bougainvillea

LOUIS-ANTOINE DE BOUGAINVILLE

Family: Nyctaginaceae, the four-o'clock family
Number of species: *c.* 10–15
Distribution: southern South America

When I first visited the Fairy Lake Botanical Garden in Shenzhen, China, my breath was taken away by the festoons of brilliantly coloured bougainvilleas cascading from the buildings at the entrance. Bougainvillea is the official floral emblem of Shenzhen and of many other cities in the tropics and subtropics worldwide like Guam, Grenada and Guangzhou – all celebrate this plant with its brilliant magenta, orange or yellow 'flowers'. They arrived in India via British colonizers and are today a significant part of Indian landscapes, with many local varieties and cultivars adorning gardens from West Bengal to Tamil Nadu. All species of bougainvillea though are native to the dry and semi-dry forests of South America, and not all species are brilliantly coloured. In the dry forests of Bolivia, I once collected one with green bracts – hardly showy at all!

The brilliant colours of bougainvillea come not from the flowers, but from leafy bracts that subtend the groups of usually white or cream-coloured tubular flowers. One of the common names for bougainvilleas is 'paper flower' from the papery texture of these modified leaves. Most plants have a flower with a whorl of sepals, the calyx, and a whorl of petals, the corolla – think of a buttercup with the yellow petals and the green sepals. Members of the four-o'clock family are a bit different. Flowers of bougainvilleas have only a single undifferentiated petal-like whorl – called the perianth. At the base of the narrow perianth tube is a large structure that secretes sugary nectar – a reward for pollinators visiting the flowers and carrying pollen to other individuals of the same species. In the cultivated bougainvilleas it is not the flowers that act as an attractant for visiting insects and birds but rather the brilliantly coloured bracts. Each flower has at its base one of these papery bracts – in the cultivated species they are a bewildering range of colours, from

The bracts of *Bougainvillea stipitata* from northern Argentina are not as showy as those of the cultivated species, but they are similarly arranged around a group of tubular flowers.

Louis Antoine, Comte de Bougainville (1729–1811) evocatively described the peoples of the Pacific, influencing later philosophers like Jean-Jacques Rousseau's ideas of the noble savage.

bright red to white, but the 'classic' bougainvillea colour is magenta, the hottest of hot pinks. Many plants have brightly coloured flower parts, but the colours in bougainvilleas and their relatives almost fluoresce they are so vivid. This is due to the pigments produced in the cells of the plants. Bougainvilleas are members of the plant order Caryophyllales, a group that has been recognized for a long time, one of whose characteristics is the production of pigments called betalains, rather than the more ubiquitous anthocyanins, that give plant parts their red colours.

Betalains – the name comes from the beet genus *Beta*, also a member of the Caryophyllales – are found in some fungi – like the fly agaric, *Amanita muscaria*, and a bacterium (*Glucoacetobacter*) – but in flowering plants only in the members of the Caryophyllales. This order of flowering plants contains some forty families – some familiar ones are the cacti, the beetroots, and of course, the four-o'clocks, of which bougainvillea is a member. In most other flowering plants, red and blue colours are the result of pigments called anthocyanins, and yellows from flavones. Both the anthocyanins and betalains are derived from amino acid precursors – anthocyanins from phenylalanine, betalains from tyrosine. Plants that produce betalains never have anthocyanins and vice versa – they are mutually exclusive.

But plants like bougainvilleas that have betalain pigments still retain the chemical pathway for producing anthocyanins, some of the early steps in the pathway are shared. Initially the two pigment classes existed together, but since they compete for the same substrate, competition can go either way – if the chemical reactions go along one pathway, anthocyanins are produced, while if the reactions go the other way, we get the brilliant betalains. The adaptive significance of these pigments is still not known, but once developed in the ancestors of a lineage, descendants inherit this 'shared derived character'. Shared derived characters define groups – and although we often cite the presence of betalains as such a character, it is really the chemical pathway itself that matters. Novel characteristics like these striking pigments can also be lost in lineages – this is the challenge of systematics. Does a pattern where some plant families in otherwise betalain-containing lineages produce anthocyanins mean that the ability to produce betalains has been lost, or has that ability evolved multiple times? Much remains to be elucidated about the chemistry and biology of these pigments that give us the bright bracts of bougainvilleas that so adorn tropical and subtropical gardens.

The genus *Bougainvillea* takes its name from the great French soldier and 18th century explorer Louis-Antoine de Bougainville. Bougainville was an extraordinary man, today remembered – at least in the English-speaking world – through the plant that bears his name. He was from a family of scholars and lawyers, and as a young man he achieved recognition for a mathematical treatise on the analysis of the infinitely small, perhaps indicating a brilliant career as an academic. Instead, he joined the military and found himself in what is now Canada, fighting the British for possession of the interior of North America in what is known as the Seven Years War. Appointed aide-de-camp to General Louis-Joseph de Montcalm, he was sent along the contested borders inspecting troops, fortifications and, in general, French readiness for conflict.

While in Canada Bougainville made notes on the land, its peoples and their customs. Montcalm wrote to the Minister of War saying "I never lose an opportunity of informing myself about this little-known country. M. de Bougainville, with whose powers of discernment you are no doubt familiar, works even harder than I do to complete this subject, and perhaps one day we may prove of use to this colony." Bougainville himself was enchanted with Canada, "What a colony! What a people!" Despite the catastrophic losses incurred in the Seven Years War, Bougainville's talents were highly esteemed by those back in France – he was hailed as a hero by the court of Louis XV and his consort Madame de Pompadour. He could have spent his life at court, but instead re-entered military service. In 1762, disgusted with the Treaty of Fontainebleau that ceded to Britain the French 'possessions' in India and eastern North America and the secret 'gift' of Louisiana to Spain, he set about an attempt to reassert French dominance somewhere. He focused on the Isle of St. Malo (the Malvinas or Falklands today) from where he made forays to colonize the as-yet uncharted islands of the Pacific. Finding the government coffers empty, Bougainville offered to finance the naval expedition himself by floating a company to which wealthy shipowners and others subscribed. Thwarted by the British, this attempt ended in failure. But the rush to the Pacific had begun, and in 1766 Bougainville set out from the port of Nantes to circumnavigate the world with two ships, the *Bondeuse* and the *Etoile*.

Plant breeders have taken advantage of natural variation in bract colour in *Bougainvillea* and created a huge array of shades ranging from yellows and oranges to hot pinks and reds.

Along with the usual compliment of sailors the expedition had an official naturalist, Philibert de Commerson. The Bougainville expedition, as the voyage is now known, was the first to have a professional naturalist paid for by royalty as part of the team – a practice

Bougainvillea   Bougainvillea spectabilis Willd.

then copied by other European monarchs setting out to colonize the world outside. Commerson also brought along an assistant, a young man named Jean Baré (or Baret), who was an integral part of Commerson's collecting and documentation of the natural history of the extraordinary places they saw. In June of 1767, the *Etoile* and *Bondeuse* landed in Rio de Janeiro, to a rather cold reception by the Portuguese authorities. Excursions outside the city limits were not permitted, surely annoying to Commerson and Baret who were seeing the riches of the tropical flora for the first time. Despite only a brief time in Rio de Janeiro, we know that the naturalists did go beyond the city limits to collect – the herbarium specimens are evidence of that! One of these was a thorny vine with brilliant magenta-red 'flowers' of which several specimens are still preserved in the herbarium of the Museum d'Histoire Naturelle in Paris. The labels on some show how Commerson and Baret were attempting to classify the plants they found in Latin, the language of science at the time. Commerson wrote (in Latin) on one of them: "[We] give to the newly founded genus a new name and derive it from the most-esteemed M de Bougainville….. someone who appreciates all fields of natural history, the arts and sciences."

In proposing to name the plant for Bougainville, Commerson and Baret were acknowledging not only his importance as the leader of the expedition, but also his interest in science. But the plant was never described by Commerson – it remained for the later botanist Antoine de Jussieu to make the name formal. Names are not able to be used until they are published, and Commerson never lived to see his collections arrive in Paris at the end of that long and eventful voyage. And the voyage was indeed eventful. After the brief stop in Brazil where *Bougainvillea* was collected, the ships rounded the tip of South America; Bougainville wrote "… anything more frightful would be difficult to conceive." Safely into the Pacific Ocean, which must have seemed peaceful after the storms of Tierra del Fuego, the ships set sail across unknown seas.

They were the first European voyagers to land in Tahiti, where they spent months in what to them was a tropical paradise. There, the disguise as a young cabin boy cum botanical assistant that Jean Baret had maintained through months of sea voyage was unmasked; Jean was really Jeanne – a woman! How ironic for Commerson, who had priggishly written about virtue earlier in his career, to have been found to have smuggled his lover aboard as his assistant. Bougainville took it in his stride and said of Jeanne, "She knew from the start that the goal was in prospect, and the idea of such a voyage had excited her curiosity. She would be the first woman to accomplish it and I must do her the justice to acknowledge that her behaviour all the time she was on board was a model of propriety." To the adventurous Bougainville it must have seemed quite logical that someone would want to come along, however unusual that might be.

The papery brightly coloured bracts of *Bougainvillea spectabilis* enclose an inflorescence of flowers whose colour contrasts with that of the bracts, orienting pollinators to where to search for nectar.

Jeanne Baret, who joined Bougainville's circumnavigation disguised as a cabin boy, was an accomplished herbalist whose botanical knowledge certainly helped Philibert de Commerson, the ship's official botanist.

Leaving Tahiti, the ships sailed for several months without making landfall – they circumnavigated the New Hebrides, proving once and for all that the continent of 'Espiritu Santo' proposed by the Portuguese for these land masses was in fact an archipelago – but supplies of food and water ran low. Half the crew had scurvy and were unfit for duty, so rather than turning to the south to what Bougainville called 'New Holland', they turned towards the Moluccas, where revictualing could take place in settlements controlled by the Dutch East India Company. The ships were repaired, new provisions were taken aboard for the journey back to France. In Mauritius (to them known as Île-de-France), Commerson and Baret left the company – remaining to study the flora and fauna of that island and of nearby Madagascar. Commerson later died in Mauritius; he never published any of the extraordinary observations he had made in the circumnavigation – it was all left for others. His name lives on in the scientific names coined for the many new species that were collected and amazingly made it all the way back to Paris at the end of the voyage. Jeanne Baret married a local French officer, Jean Duberna, and returned to France, where she was awarded a pension by the French government in recognition of her bravery and contribution, something supported by Bougainville. The genus Commerson proposed to name after her, *Baretia*, was never published. Bougainville – again – returned to France a hero. Despite the lack of new territories for the French crown, he had circumnavigated the globe, and his popular account of the voyage was almost universally acclaimed. He was still thirsty for adventure though, proposing an expedition to find the north-west passage through the Arctic, but due to lack of French government funds this was never undertaken.

All his life Bougainville valued merit over social station; he railed against the custom of the day to promote men for their titles rather than their deeds. So, it might have been expected that he sympathized with the revolutionaries at home in France, but he remained a staunch defender of the royal family and was at Louis XVI's side as mobs broke into the Tuileries. Amazingly, he survived the Terror, re-emerging as one of the architects of Napoleon's military successes. He died acknowledged as a great explorer and citizen of France and is buried in the Pantheon. The plant named for him, *Bougainvillea*, lives on in gardens and greenhouses worldwide, surely an even more fitting memorial to his contribution to our knowledge of the natural world.

Commelina

JAN COMMELIN

Family: Commelinaceae, the dayflower family
Number of species: 150–200
Distribution: cosmopolitan, mainly tropical

In his *Critica Botanica,* Linnaeus explained why the name *Commelina* was appropriate for the dayflowers, he wrote "*Commelina* with three-petaled flowers; two petals beautiful and a third inconspicuous, for the two Commelins and a third who died before accomplishing anything in botany [translated from the original Latin]." He took the generic name from a previous work by the French botanist Charles Plumier, who had indeed dedicated the genus to the Commelins, Jan and his nephew Caspar. Linnaeus embellished this original simple dedication with his own romantic reference to the number of petals and to a son of Caspar's. Neither of Caspar Commelin's sons were botanists at all, one died in infancy and the other was a medical doctor of some note in Amsterdam who was only three years old when Plumier's book was published.

Plumier's dedication was more clearly focused, if slightly less romantically, on the botanical achievements of the Commelins, "For the famous Commelins, Jan Commelin, senator of the city of Amsterdam (while he was alive), and Caspar Commelin M.D. of the Hortus Medicus and physician of Amsterdam, where were cultivated rare plants from the East and West Indies etc., for which there were descriptions and drawings of living plants made." (my translation from the original Latin).

Jan Commelin was an important merchant in pharmaceuticals, supplying pharmacies and hospitals in Amsterdam and throughout the Netherlands in the mid-17th century. He held political office in the city of Amsterdam, eventually becoming one of the thirty-six Councillors who ran the city, a position

Jan Commelin (1629–1692) amassed his fortune and status through his trade in medicines, then derived mostly from plants, and was powerful in the city government of Amsterdam.

he held for more than twenty years. Merchants like Commelin were at the centre of society in this golden age of Dutch expansion. The Dutch East India Company (Vereenigde Oost Indische Compagnie or VOC) had formed from the amalgamation of several independent trading companies in the early 17th century, first importing silks. Once the Dutch government granted the VOC a monopoly on the spice trade, it grew to become what has been called a 'proto-conglomerate', or the first truly transnational corporation, the organisers of a global supply chain, trading and producing spices mostly from the islands of what is now Indonesia. They also invested in shipbuilding and industry – the VOC was a model of capitalist enterprise.

Building on the botanical expertise he acquired as part of his trade and based on his great interest in plants, Jan Commelin began publishing botanical works, beginning with a book on the cultivation of citrus in the Netherlands. Shortly after he entered the city government, he bought a property south of Haarlem and developed a collection of exotic plants, almost certainly obtained from VOC voyages. He edited and annotated all but the first volume of the *Hortus Malabaricus*, a catalogue of the plants of Cochin, the VOC headquarters in Malabar (today's Kochi in the state of Kerala, India) compiled by Hendrik Adriaan van Rheede tot Drakenstein, the Dutch governor of Malabar. Rheede tot Drakestein obtained his information from Itty Achuthan, a local practitioner of traditional medicine, who dictated in Malayalam from manuscripts written on palm leaves. Later the medical information was translated to Latin, for editing by Jan Commelin, among others. The work focused on plants of medical importance, surely of great interest to Commelin in his trade in pharmaceuticals, which were at the time almost entirely derived from plants.

In 1682 the Amsterdam city council voted to establish a 'Hortus Medicus' – a garden devoted to medicinal plants in the area of De Plantage on the outskirts of the city. Jan Commelin and Joan Huydecoper van Maarseveen, both councillors, were appointed as commissioners of the new establishment. The Hortus Botanicus of Amsterdam is still on this site today. Jan Commelin began the monumental task of getting illustrations created in colour of all the plants grown in the Hortus Medicus, along with the publication of a catalogue of all the plants grown there. His responsibilities kept growing; he was appointed 'forister' for the city of Amsterdam giving him authority over the Hortus as well as all the other city-run plantations, and finally in 1690 he was paid for his efforts. When he died in 1692 only one volume of his catalogue had been published; it was the next generation who completed the work.

In 1696, after obtaining a medical degree, Jan's nephew Caspar was appointed the botanist of the Hortus Medicus. He followed in his uncle's footsteps, first publishing an index to the *Hortus Malabaricus*, in which he listed those names alongside all other names given for the same plants. He carried on with the project of providing colour

Commelina communis is one of the many host plants for the larvae of this moth
– described in the 18th century as the wild comfrey moth, after the common
name of the plant at the time in the state of Georgia, USA.

illustrations for all the plants cultivated in the garden, commissioning paintings from a variety of artists, including some from the daughter of Maria Sybilla Merian (see *Meriania*), Joanna Helena Herolt. Over the years, Caspar and Joanna developed a close working relationship – eventually he provided the botanical annotations for Merian's book on the metamorphosis of Surinamese insects. Caspar published several catalogues of the plants in the garden, but his publishing in botany ceased as he obtained postions first as the physician for the VOC (for which he was paid in an annual share of spices!) and later as the inspector of all medical and pharmaceutical activities in Amsterdam. Medicine was botany in the 17th century, and the Commelins, with their mercantile and medical interests, had significant impact on the later development of botany as a profession of its own.

Those two showy petals that for Linnaeus symbolized Jan and Caspar last only a few hours, not even an entire day, hence the common name dayflower. *Commelina* species are herbaceous plants with grass-like leaves, whose flowers are held in groups in two enclosing bracts, called a spathe. Each flower has two large, upwards-pointing, coloured petals, often either blue or orange depending on the species, and one much smaller white petal that points down, plus an unusual arrangement of six stamens, or pollen-bearing organs. But not all of these stamen-like structures are actually full of pollen – they can be divided into three types. Three bright yellow cross-shaped structures are on short, straight stalks (filaments) and are central in the flower – these are staminodes, with no or minimal pollen, but with an important function in pollinator attraction. Two stamens on long filaments arch down on either side of the lower white petal and style – their anthers (called the lateral anthers) are full of pollen, but are dull in colour and not very conspicuous. The last type of stamen is presented together with the three staminodes on an upwards-curving filament; its anther (called the medial anther) is bright yellow and is full of pollen – this acts as a reward for the flies and bees that visit the flowers, *Commelina* flowers have no nectar. Experimental manipulation of *Commelina communis* flowers showed that the bright blue petals attracted pollinators from afar – mostly flies of the family Syrphidae, the hoverflies – to flowers. Syrphid flies are hovering flies, striped yellow and brown, looking a bit like delicate bees, and like other flies are important pollinators of many flowering plants.

The bright yellow staminodes and pollen-filled medial anther serve to orient the insect on the flower so their abdomen touches the lateral anthers, dusting them with pollen to carry to the next flower where it is brushed onto the stigma and grows down to fertilize the ovules to produce seed. Without the staminodes the insect hovers for longer before trying to enter the flower and often just gives up – the cues are not there. The medial anther does not deposit pollen on the insect body but rather is fed upon by visiting insects, functioning to position the insect correctly in the flower so they brush against the lateral anthers to pick up pollen. The intricate interplay between the various parts of *Commelina* flowers provides a set of guides, orienting a pollinator in the flower to maximize pollen transfer from flower to flower.

Hoverflies don't land on commelina flowers with the bright yellow X-shaped staminodes removed, showing these sterile structures are important for attracting these key pollinators.

Some *Commelina* species produce not only showy flowers with brightly coloured petals, but also have what are called cleistogamous flowers that are borne underground near the base of the plant. These tiny flowers never open and are entirely self-pollinated; insurance against failure of pollination of the showy ones out in the open. Some of these *Commelina* species are weedy and invasive; they are very shade tolerant, resistant to drying out, persist for a very long time in the seedbank, and can choke out other vegetation by their quick growth and spread. Of the nine species of *Commelina* in North America, six of them are introduced, with one, *Commelina benghalensis*, being listed as a noxious weed due to its adverse impacts on cotton, peanut and soybean crops in the southern USA. In China *Commelina communis* has been documented to cause significant reduction in soybean yields, and in the Caribbean *Commelina diffusa*, once planted as a ground cover to reduce erosion, is a host of a number of significant nematode-borne diseases of banana. Control is difficult, and the advent of new low-tillage or no-herbicide agricultural systems has increased the problem for banana farmers in the tropics as populations of the weed expand unchecked.

A significant new problem associated with *Commelina* in agricultural fields is herbicide resistance. Herbicides, applied to fields to kill weeds but leave resistant crops unaffected, were long used as potent control measures in agriculture. Today, however, there are 266 species of flowering plants identified as weeds that have evolved herbicide resistance, not just to one herbicide, but to 164 of the herbicides used by farmers. Some of these are *Commelina* species. Of course, not all *Commelina* are nasty invasive pests, many are native plants – but the invasives are spreading fast, aided by the introduction of minimal-tillage systems with herbicide-resistant crops, created by genetic modification, often for glyphosate resistance. It seems ideal, a crop that can resist damage by a potent herbicide, less spraying of the chemical, thus less environmental damage. But evolution is real, and some weeds bounce back; here neither the weeds nor the crops are harmed by herbicides. In fact, for *Commelina communis* in the United States, it seems that repeated use of glyphosate reduces the competition it faces from other weeds, thus allowing it to run rampant. These plants have evolved tolerance to the herbicide, not unexpected as herbicidal application results in very strong selection pressure on the populations of weeds, leading to adaptive change favouring survival.

These invasive *Commelina* species have adapted to resist many standard herbicides; the problem is not just with glyphosate, a great many other chemicals and chemical cocktails have limited effects on these resilient plants.

So, if herbicides are no longer effective for the control of invasive, weedy *Commelina* species, what can farmers do? Mechanical control has been suggested as the only way forward, but this too has its problems. Many of these invasive species root easily from broken stems, so scything or 'weed-whacking' just spreads the plant further. Even tiny pieces of stem can regenerate to form new plants – it's impossible to get every last one except in very small plots. Those pale underground cleistogamous flowers also can produce hundreds of seeds, out of sight. Biological control by herbivorous insects or fungal and bacterial pathogens has not yet been successful. More integrated methods for control have been suggested for smallholders; mulching with organic waste, interplanting with another shade-tolerant crop like melons, sweet potatoes, or cowpeas that can outcompete the aggressive invasive dayflowers all show promise on that small scale. In some parts of the world commelinas, characterized as noxious weeds of modern agriculture, are eaten as leaf vegetables by people and used as valuable fodder for livestock.

We should not be surprised that plants have adapted to our high intensity use of chemicals to control their spread where we do not want them. This is how evolution by natural selection works; only those individuals who survive to reproduce contribute to the next generation. This is how the diversity of life on our planet has arisen; we should not be surprised if the process doesn't always follow our human rules or desires.

The inflorescence of all *Commelina* species is cupped by protective bracts; a single flower opens per day and lasts only a few hours.

Darwinia

ERASMUS DARWIN

Family: Myrtaceae, the myrtle family
Number of species: *c.* 50–70
Distribution: southern Australia

You might be forgiven for thinking that any plant named for a Darwin was to honour Charles Robert Darwin, whose ground-breaking book *On the Origin of Species by Means of Natural Selection* published in 1858 truly changed the way we think about the world and our place in it. But in fact, this beautiful genus of Australian shrubs was named from Charles's grandfather Erasmus, an English physician and one of the lynchpins of the 'Midlands Enlightenment'. Edward Rudge dedicated the plant to "the late Erasmus Darwin, M.D., of Litchfield, Author of The Botanic Garden, Zoonomia, and a translation of the Systema Vegetabilium of Linnaeus." Rudge was also from the Midlands intelligentsia; he owned an estate at Evesham and was appointed High Sheriff of the county of Worcestershire in 1829. It seems likely that Rudge and Darwin were acquainted through various learned societies such as the Linnean and the Royal Society.

Erasmus Darwin was a larger than life 18th century character, even for that time of polymaths. Educated in the classics at Cambridge and in medicine at Edinburgh, his medical practice in Lichfield just north of Birmingham in the English Midlands was his main source of income, but his interests were amazingly diverse. He was a driving force behind the Lunar Society, a group of scientist-philosophers whose meetings were held each month on the night of the full moon, allegedly so members could walk home safely. The Lunar circle was attended and promoted by such scientific and engineering luminaries as James Watt, the inventor of the steam

Original specimen used to describe *Darwinia fascicularis*, collected by Joseph Banks and Daniel Solander in April to May 1770 around Botany Bay, Australia.

Frederick Polydore Nodder completed paintings begun by Sydney Parkinson, the artist aboard HMS *Endeavour*, of the plants collected by Banks and Solander – Parkinson only completed the barest sketch of *Darwinia* before he died on the voyage.

engine and his business partner the manufacturer Matthew Boulton, Joseph Priestley, who elucidated the nature of oxygen, and Josiah Wedgwood of pottery fame. Benjamin Franklin even came to visit the group in the mid-1700s and returned to conduct experiments on electricity with some of its members. The Lunar Society was far from a provincial set of friends chatting about science – it was a hotbed of invention and innovation and considered so by all at the time.

As a country doctor, Darwin's work was heavy and involved much travelling over a large area – his patients included the rich and poor alike, the former he charged, while the latter he treated for free. Medicine in the 18th century was a pretty brutal affair and treatments were often as bad as the illnesses themselves; involving many repeat visits and much trial and error. Like his friend Josiah Wedgwood, he was an ardent abolitionist, considering the enslavement of human beings an abomination while paradoxically aiding and abetting the industrial mechanization of manufacturing that put so many people into abject poverty in England.

Like others in the Lunar Society Darwin was a prolific inventor and engineer. He had interests in canal building and navigation

Erasmus Darwin (1731–1802) was a larger-than-life figure who was dedicated to his local area. He turned down an invitation to become private physician to George III to stay in the English Midlands.

and was deeply involved in the design and building of the canal system around Lichfield and Trent. His inventions ranged from vehicle design (his attempts to alleviate the discomfort of his endless carriage journeys for his medical rounds led to the design of what is now the steering mechanism for automobiles), to speaking machines that mimicked the human voice, to ingenious mechanisms for copying letters automatically, to proto-air conditioners – all meticulously recorded in his 'Commonplace Book' with drawings and comments. In addition to engineering he shared the fascination of the age with chemistry, and later in life worked out the principal of photosynthesis, where plants convert carbon dioxide (called by him carbolic acid) and water to oxygen, powered by sunlight. In his last book *Phytologia* he also clearly articulated to the principles of plant nutrition, citing nitrogen and phosphorus, along with calcium, as critical for plant growth.

All his life Erasmus Darwin was fascinated by life – through his profession, but also through his interest in the natural world. The Swedish botanist Carl Linnaeus had revolutionized and shocked the botanical world with his classification of plants using an entirely new system based on numbers of male and female parts in the flowers. In his presentation of this "sexual system" he analogized the flower to the marriage bed, with differing numbers of brides and grooms. As you can imagine this was too much for some and was likened to "loathsome harlotry" by one of his contemporaries. Linnaean ideas became popular in English botany, and Darwin formed 'A Botanical Society, at Lichfield' in order to translate and promulgate Linnaean botanical knowledge. The Society (but really only Darwin himself, the only other members were the linguist Sir Brook Bootheby and John Jackson, proctor of Lichfield Cathedral) translated two of Linnaeus's works and published them as *A System of Vegetables* (from *Systema Vegetabilium*) and *The Families of Plants* (from *Genera Plantarum*) in which many of the English common names we use today were coined. Although now in English and therefore accessible to those not knowledgeable in Latin, these books were still pretty tough going, dense and quite frankly a bit dull.

Darwin went one step further to bring Linnaean ideas to wider English-speaking public though with a botanical poem, *The Loves of the Plants*, which was the first part of his larger work *The Botanic Garden*. With this he hoped to interest readers in the "delightful science" of botany through 'Imagination under the banner of Science'. First published anonymously in 1787 and later under his own name in 1789, Darwin takes the reader through the twenty-four classes of the Linnaean sexual system, all based on the numbers of male (stamens) and female (pistils) parts in the flower. It drips with sexual innuendo and its mock-heroic tone echoes the poetry of the day; he humanises plants, creating word pictures which are then explained with copious footnotes explaining just what is going on. Describing the dyer's broom, with its ten stamens and one pistil:

"Sweet blooms GENISTA in the myrtle shade,

And *ten* fond brothers woo the haughty maid."

The footnotes to the poetic cantos are long and explanatory, revealing deep understanding not only of the knowledge of others, but of plants themselves and their ecology. Throughout the poem it is assumed that the reader has knowledge of classical literature, and of current events – the connections are astonishing. The plumes of the thistle are likened to Montgolfier balloons, and the image of "bloated Dropsy pants behind unseen" illustrates the perils of overindulgence in wine – the product of *Vitis*. Darwin's verses were inspiration for many of the romantic poets and echoes of his phrases can be found in the works of Wordsworth, Shelley, Coleridge and Keats, although they later mocked his poetic style.

I wonder what he would have thought of *Darwinia*, with its groups of flowers each with a single style and ten stamens and ten staminodes (sterile anthers, with no pollen, called 'eunuchs' by Linnaeus and Darwin!). The styles are long and grow out as the flowers open, pushing through the pollen on the stamens as they elongate – I can just imagine what couplets might come from that! Linnaeus never saw plants from Australia, but Darwin probably did, through his acquaintance with Joseph Banks, whose specimens Rudge used to describe *Darwinia*, but neither Rudge nor Darwin would have seen live plants.

The common name for *Darwinia* species is 'mountain bells' or 'bells' due to the enlarged brightly coloured usually leafy bracts that surround the groups of flowers. The genus is only found in Australia and although at first glance doesn't look like it, is distantly related to eucalyptus and bottlebrush trees. Most of the species occur in

Western Australia, where new species are still being described; most of these are rare and endangered – they occur in tiny areas of very specialized habitat, but threatened not only by human landscape alteration but by climate change.

Populations of different *Darwinia* species hybridize where they come together both in eastern Australia, where Banks and Solander first encountered these plants, and in the drier habitats of the west. Hybridization, where two different species cross and form fertile offspring, was a problem for those who considered that all organisms had been placed on Earth by a

The inflorescence of this *Darwinia taxifolia* is composed of several flowers grouped together and surrounded by brightly coloured bracts such that it looks like a single flower.

Creator and had not changed since. But it clearly happens in the wild as well as in the greenhouse and not only results in sterile 'mules' but in fertile plants.

So why are there still different recognisable entities we call species in nature? Why don't they all blend together? Many evolutionary biologists have pondered this question for many years, including Erasmus's grandson Charles Darwin. Erasmus himself saw the interconnectivity of life – in his great work *Zoonomia; or, The Laws of Organic Life* he clearly articulates the commonality of all living creatures and concludes:

> "Would it be too bold to imagine, [...], that all warm-blooded animals have arisen from one living filament... thus ... delivering down those improvements by generation in its posterity, world without end!"

Not a question, but the assertion of common descent of all animals and by analogy all of life – in one place he even says "Go, proud reasoner, and call the worm thy sister!" It fell to his grandson to articulate the mechanism by which change happens – natural selection. And it is at least in part natural selection that allows for species to remain distinct even though they can hybridize; in fact, some hybridization can help with the exchange of advantageous genes between different species.

Some *Darwinia* species are so specific in their requirements as to be confined to single populations in tiny localities; these are in danger of extinction. Just as natural selection is involved in the origin of species, it too is involved in their ultimate demise, usually helped along by human expansion and alteration of the environment in which they have existed since long before humans came on the scene.

Eastwoodia

ALICE EASTWOOD

Family: Compositae, the daisy family
Number of species: 1
Distribution: California, USA

In dedicating the genus *Eastwoodia* and its only species *Eastwoodia elegans*, Townshend S. Brandegee merely stated "Named in honour of Miss Alice Eastwood, curator of the herbarium of the California Academy of Sciences." Little did he know at the time the important role Alice Eastwood would play in the understanding and discovery of the flora of California.

In 1894, when *Eastwoodia* was first described, Alice Eastwood had only just come to the California Academy of Sciences from Colorado; the Brandegees, Townshend and his wife Katharine, also a botanist, had convinced her to join them in San Francisco. They had been impressed with her knowledge of botany when she visited them a few years before on a collecting trip. She stayed at the Academy in San Francisco for the next 55 years – building it into a premier collection of Californian plants, twice. Eastwood was a self-taught botanist. From her teenage years she lived in Colorado, where she took on several jobs to help her widowed father make ends meet. Despite the hardship in which the family lived, she graduated valedictorian of her class. Her trips to the Colorado mountains cemented her love of nature; field trips to the mountains were a joy. However, she needed to earn a living, so took a job as a teacher in her old high school – scrimping and saving to fund her summers of field work in the Rocky Mountains. She usually travelled alone and on horseback; she pioneered the use of practical riding gear for field work – she rode astride rather than side-saddle, a novelty at the time. She had adventure aplenty, her ambition to scale Gray's Peak (4,350 m/14,270 ft) got her lost, her money stolen, but she was unfazed – "As a matter of fact, I always had more concern for my plants than for my money."

When the eminent British natural historian Alfred Russel Wallace came through Colorado in 1887 on his American lecture tour, who better to lead him to the summit of Gray's Peak but the young Alice Eastwood. The two, despite the forty-year gap in their ages, became fast friends after their botanising of Colorado's beautiful alpine flora, and Alice visited Wallace and his family in England later in life.

By the last decade of the 19th century Eastwood had been persuaded by the Brandegees to re-locate to California, where Katharine offered Alice her own salary to act as joint curator of the herbarium at the California Academy of Sciences with her. By 1893, Eastwood was the sole curator, the Brandegees had moved to San

Diego, taking with them their library and private herbarium. Alice set to work to care for the collections that remained and, more than that, augment them with more collections of her own. She collected plants all over California, alone or with interested colleagues never minding hardship. Her knowledge of plants became respected throughout the male-dominated learned circles of American botany. She exchanged duplicate specimens with the doyen of American botany of the time, Asa Gray of Harvard University, one of which was the specimen she collected on 10 May 1893 near the town of Alcalde that was described as *Eastwoodia elegans*.

Alice Eastwood (1851–1953) was born in Canada but devoted her life to the study and protection of the flora of California; she fought tirelessly to conserve areas of special botanical interest.

Always unconventional, Eastwood decided to keep the type specimens – those used in the description of new species or genera – separately from the rest of the herbarium at the Academy. Usual practice was, and still is in many collections, to intercalate these precious specimens with others of the same taxon, allowing ready comparison and checking. It turned out that Alice's decision was an incredibly good one.

In the early hours of the 18 April 1906, all of San Francisco awoke to the rumble and shake of a magnitude 7.9 earthquake. The newly constructed building of the Academy was in ruins, the collections at risk from not only the collapse of the structure but from the fires that raged throughout the city. Awakened by the quake, Eastwood rushed to the Academy, leaving her own possessions and home behind. There she found the botanical collections on the sixth floor completely inaccessible. Ascending the collapsed marble staircase by the iron railings, she reached the cabinets where the type specimens were stored and using a system of ropes and pulleys lowered the boxes of precious plants to the ground where colleagues were ready to load them into a cart to take the type specimens out of harm's way. These 1,500 precious specimens were driven around the city the rest of the day, avoiding the fires that sprang up throughout the area in the wake of the earthquake. Alice wrote "Nobody knew where the safe place would be; for it seemed the whole city must go…." Other parts of the collection and archives were saved by other curators, but by the end of the day most of collections at the Academy had been destroyed. The only personal item Eastwood saved from destruction was her lens – the botanist's indispensable companion. Three weeks after the earthquake Alice wrote:

This herbarium sheet of *Eastwoodia elegans* was collected by Alice in the Maricopa Hills in 1913 as part of her rebuilding of the collections of the California Academy after the 1906 earthquake.

"I do not feel the loss to be mine, but it is a great loss to the scientific world and an irreparable loss to California. My own destroyed work I do not lament, for it was a joy to me while I did it, and I can have the same joy in starting it again. ... All my pictures and books are gone and many treasures that I prized highly; but I regret nothing for I am rich in friends and things seem of little account. ... I am beginning to recollect and intend to go to type localities as much as possible, I expect the academy will be able to give me but little aid for the present, but have a tiny income of my own and can get along, I feel sure."

For the next six years Eastwood travelled to visit herbaria in the eastern United States and Europe, funding her travels through her own meagre income from rental property in Colorado. There were no funds from the Academy to rebuild, so she set about it herself, by collecting again many of the plants that were lost. She attempted to find employment at the Smithsonian, but to her shock discovered that because she had been born in Canada, she was a British, not an American, citizen; she immediately set about remedying the situation, but was not granted citizenship until 1918. In Europe she visited collections in Cambridge, London and Paris, studying specimens collected on earlier voyages to the western United States and working on botanical treatments of Californian plants. The friendships she made during these travels were lifelong; her dedication to science and the depth and breadth of her knowledge were esteemed wherever she went.

Returning from Europe she found an invitation from the Trustees of the Academy in San Francisco to re-join the staff with the offer of some funds to begin to equip the herbarium. The Academy itself was in temporary accommodation, but a decision had been made to re-build on the site of Golden Gate Park and to accompany the scientific centre with plantings of plants from all over the world. Her enthusiasm for creating a botanical garden was stimulated by conversations with the German botanist Adolf Engler, who, with other European enthusiasts for plants and ecology, had come to California on the last leg of their journey across the country to learn about the landscapes of the United States.

In the summer of 1914, she went on an extended collecting trip to the Yukon to collect Arctic willows for Charles Sprague Sargent of the Arnold Arboretum.

Managing to avoid customs duty on her many bags of specimens by pointing out that they were also intended for the herbarium of the Canadian government in Ottawa, she returned to San Francisco to discover that the new building for the Academy did not include space for a botany department. She soon set that right with her usual directness and determination. She continued to collect all over California and to help plant-lovers, professionals and amateurs alike, using the gardens to promote interest in the cultivation of Californian plants. In collecting she worked closely with John Thomas Howell, who drove everywhere once she was unable to do so. With Howell she collected near the car, he ranged further afield – she wrote to a friend "My days of exploring on foot are over, but one can do a good deal from autos, supplemented by one's legs." The indominable spirit of those early days on horseback never died. Eastwood collected well into her 90s, and by the end of her life had added almost 350,000 specimens to the collections of the Academy's herbarium, her "child ...dearer to me than life." She had rebuilt the collections from scratch, and they are still essential for understanding the rich and varied California flora.

California's plant life is extraordinarily diverse and rich in endemics, species that grow nowhere else. California, the state and adjacent areas also known as the California Floristic Province, has long been considered one of North America's biodiversity hotspots, and of the more than 5,000 flowering plant species known there, around 40% of these are endemic. The great size and varied topography of the region contributes to this pattern; low extinction, rather than high speciation, rates have been suggested as causes for this high diversity. Many lineages of plants have diversified in the arid habitats that are part of the Californian landscape. Among these are the composites or members of the daisy family, of which the genus *Eastwoodia* is one. *Eastwoodia* belongs to a large and incredibly diverse North American radiation of the tribe Astereae – the asters and their relatives – of the megadiverse family Compositae.

The capitulae (heads) of *Eastwoodia elegans* are composed only of disc flowers, unlike those of dandelions, which are composed entirely of ray flowers.

Composites take their name from the many flowers that form a tight, flower-like inflorescence – what looks like a flower in a daisy or a dandelion is really a group of flowers tightly packed together. Some 'comps', as botanists affectionately call them, have what are known as ray flowers – those yellow strap-like structures of the

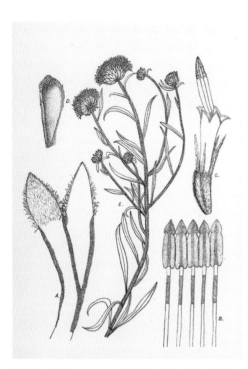

This illustration accompanied the original description of *Eastwoodia* and in the tradition of botanical illustration has close-up details of flower parts like stamens and petals.

dandelion inflorescence – while others have only disc flowers – tiny tubular structures in a head - and others have both – like daisies. The Compositae, also known as the Asteraceae, share the honour of being the largest family of flowering plants with orchids (Orchidaceae); which has more species depends upon who you are talking to. Nevertheless, with some 32,000 species, the composites comprise almost 10% of flowering plant diversity, a botanical challenge indeed. The western parts of North America are particularly rich in species of Compositae, and the family has been used as a model system with which to study the relative roles of phylogeny (evolutionary relatedness), soil geochemistry and topography in the process of diversification.

Alice Eastwood, who was well-known for the joy with which she named species for colleagues, would probably have been pleased that this genus, with its single species, has stood the test of many different data types to remain accepted today. *Eastwoodia* is related to the species-rich lineage of rabbitbrushes (*Chrysothamnus*) that has radiated in the arid American Southwest but is sufficiently distinct both morphologically and in DNA sequences to be considered a distinct branch of evolution.

It is fitting that such a distinct and distinctive plant be named for Eastwood – she was distinct and distinctive herself. During her lifetime she was the only woman to be starred as a person of "outstanding accomplishment" in the compilation called 'American Men [sic!] of Science', she was the honorary President of the International Botanical Congress of 1950 in Uppsala, Sweden – an honour conferred on her by her European colleagues in recognition of her extraordinary lifetime of achievement in botany. She was also fierce in her personality. John Thomas Howell, her close colleague and successor as curator of the Academy's herbarium said of her "Her impatience could be as violent as her kindness and generosity were great, and the force and bite of that impatience were dreaded by all who encountered it." Alice Eastwood was an extraordinary woman in a world where professional women were not the norm – she was the definition of indefatigable. At the age of 94, lying ill on her deathbed she is said to have told her friends "Of course I will get well, I've always been a healthy person."

Esterhuysenia

ELSIE ESTERHUYSEN

Family: Aizoaceae, the fig-marigold family
Number of species: 6–7
Distribution: South Africa

When we think of extreme plant diversity, the Amazon basin with its lush tropical rainforests often springs to mind. But the southernmost tip of Africa is home to the most species-rich terrestrial flora anywhere. The Cape Floristic Region has extraordinary plant diversity for its size, and most of this is found nowhere else, making the region one of the Earth's biodiversity hotspots. Much of this diversity is in the vegetation type called fynbos, which dominates the area of the western Cape in South Africa. Fynbos – from the Dutch 'fijn bosch' meaning 'fine bush' or kindling, as opposed to wood for burning that comes from trees – is a low shrubby vegetation with many variants depending upon soil type, aspect and other environmental factors. The proteas (sugarbushes), ericas (heaths), and restios that all have their centres of diversity in the fynbos are striking dominants in the vegetation; the variety is breath-taking, about 700 of the 800 species of *Erica* occur only in the Cape region, and one-third of all species of the family Restionaceae are found in the region. This high diversity has been attributed to speciation triggered by climatic change caused by the change in ocean currents, creating cooler, more arid environments, But analysis of dated phylogenetic trees of species-rich Cape lineages suggests that a combination of environmental complexity and stable climate are the overall main drivers of this hyperdiversity.

The Cape Floristic Region has been the subject of much botanical study, in part due to its extreme diversity, but also because it is seriously under threat. As a vegetation type the fynbos has been characterized as critically endangered – from increasing human population with accompanying agricultural land conversion, invasive species and, of course, climate change. As a fire-dependent vegetation – like other areas with high winter rainfall and hot, dry summers such as California, southwestern Australia and the Mediterranean region of Europe – rising temperatures create conditions where fires burn hotter and with more frequency, disrupting reproductive cycles of plants that have evolved alongside previously fire-stable regimes.

Of all the many different types of fynbos, one stands out as being less well-researched than the rest and is found high elevations in the rugged mountains. Altimontane fynbos is highly fragmented on island-like mountain tops above about 2,000 m (6,560 ft) high, and grows on thin, erosion-resistant, quartzite-rich soils

The low-nutrient sandy quartzite soil of South Africa's high mountains is a real challenge for plants, but here on Olifantsberg *Esterhuysenia alpina* succeeds where others fail.

and rocky outcrops – plants here really need to be able to hang on! In addition, these areas are usually snow-covered in the winter, and as a result, the habitat is somewhat wetter than that of other types of fynbos. It is not surprising that this type of fynbos is not well-researched, it is fiendishly difficult to access. Much of what is known about the altimontane fynbos comes from the work of one intrepid woman, Elsie Elizabeth Esterhuysen – described as a 'legend' by her colleagues. She devoted her life to the South African flora and, most particularly, to the collection and documentation of the flora of the high mountain peaks.

Elsie Esterhuysen was born in a South Africa very different from that of today. She was denied a position in the group undertaking the botanical survey of the country because she was a woman – as one of her colleagues remarked, "In those days the very prospect of a female doing botanical survey work in the remoter parts of the South African bush was unthinkable!" She certainly proved those bureaucrats wrong. In the late 1930s she began her life-long association with the Bolus Herbarium in Cape Town, at the time managed by Louisa Bolus – another extraordinary South African female botanist, who holds the record for being the woman who has described the most plant species ever! For her first 18 years at the herbarium Elsie had no steady wage and was paid out of petty cash; the curator's battle with the administration of the university eventually resulted in a permanent position in 1956, but it wasn't easy. It was Louisa Bolus who recognized plants collected by Elsie in the Hex Mountains as distinct, naming the genus *Esterhuysenia* – both the collections she cited in the original description were Esterhuysen's.

Over the course of her career Esterhuysen collected almost 40,000 specimens – an enormous number considering the difficulty of the places in which she collected. She ranks as the most prolific collector for the Cape floristic region, and in the top three for South Africa overall, all of this without transport of her own – early on she had been encouraged to learn to drive, but after getting stuck on a railway crossing, she eschewed driving herself and relied on others to get her to the mountains. Her interest and proficiency in mountaineering meant this was easy – her colleagues in the Mountain Club of South Africa were all too willing to help with forays into the field. She was so valued by members of the mountaineering community that in later life she was honoured as 'a free woman' by the local Drakensberg branch of the club; she celebrated her 80th birthday by climbing the Sneeuberg (2,000 m/6,560 ft high) in the Cederberg Wilderness with the Mountain Club. In hilly Cape Town she got everywhere by bicycle – hence her renowned ability to out-hike young undergraduates in the mountains!

Esterhuysenia is one genus she collected in the high mountains, growing as a low shrubby mat amongst rocks. The collecting notes on one of Elsie's specimens cited by Bolus reads: "Growing from crevices and depressions in massive rock formations on plateau…" *Esterhuysenia* is a member of the Aizoaceae – succulent plants with showy flowers, another of those families whose centre of diversity is in the Cape region. The family is most diverse in the habitat known as the Succulent Karoo, another of the world's biodiversity hotspots, but significant diversity occurs in the fynbos as well, especially at high elevations.

The flowers of *Esterhuysenia*, like those of other members of its family, are showy and almost daisy-like – with many narrow 'petals' surrounding a brush of yellow anthers in the centre. But these are not in fact petals at all. In daisies, the strap-like 'petals' are each an individual flower, while in *Esterhuysenia* the narrow structures in the flower are derived from sterile stamens (staminodes) and so are more properly referred to as 'petaloid', meaning petal-like. They perform the same function as petals though, attracting potential pollinators to the flowers with their bright colour. The bright pink colour of the flowers of *Esterhuysenia* gives a clue to the affinity of the Aizoaceae more generally – that shocking pink is due to betalain pigments, found in most members of the order Caryophyllales (see *Bougainvillea*) rather than the more common pink pigments derived from the chemical

At home in nature, Elsie Esterhuysen (1912–2006) had an encyclopaedic knowledge of the flora of South Africa's high mountains and was a true plant collector's collector.

Growing where its roots can reach water, this mound of *Esterhuysenia inclaudens* is putting on a real show of bright pink blooms, enough to attract any passing pollinator.

compounds called anthocyanins. Think beetroots rather than radishes.

Succulent, fleshy leaves are a characteristic of Aizoaceae, *Esterhuysenia* included. This characteristic often makes Aizoaceae prized succulent garden plants, like the stone plants (*Lithops*) whose cylindrical fleshy leaves look for the all the world like the stones amongst which they grow. Although we do not know the type of photosynthesis *Esterhuysenia* employs, many plants with succulent leaves, including many Aizoaceae, use a special form of photosynthesis called CAM (crassulacean acid metabolism) that compartmentalizes the chemical reactions of photosynthesis, so that the tiny pores on the leaf surfaces through which gas is absorbed and water lost (stomates) can be closed during the day when water loss is a risk. The stomates open at night to take up carbon dioxide for later conversion into sugars that fuel plant growth, without the water loss that would occur during the heat of the day. In CAM photosynthesis, carbon dioxide taken up through the stomates at night is converted to an acid and stored in vacuoles – hollow spaces in cells – during the hours of darkness to be then transported to the chloroplast and reconverted to carbon dioxide for use in photosynthesis once the sun comes out. This specialized form of photosynthesis allows plants to save water in environments where water can be scarce; even though the altimontane fynbos habitat of *Esterhuysenia* is somewhat moister than more lowland fynbos, the soil is so thin, rocky and sandy that water is certainly a limiting factor.

Succulent plants make terrible herbarium specimens, those wonderful fleshy three-dimensional leaves of live plants dry to shrivelled little stubs. Thus, by including field notes from living plants, presumably made by Esterhuysen in nature, Louisa Bolus made the original description of *Esterhuysenia* richer in detail and more useful to those looking at plants in their natural habitats. In this original description no mention is made of the origin of the genus name, nor any mention of the extraordinary contribution Elsie Esterhuysen made to knowledge of the Cape flora. Maybe it was just understood, or maybe, since Louisa Bolus and Elsie Esterhuysen worked together in the same institution for many years, it was an expression of Elsie's famous modesty about her own achievements. One of her favourite comments was apparently "I'm only filling in gaps." Esterhuysen rarely published any of the new species she found and collected, she left that to others – the 'academics'. She was famous amongst South

African botanists for her assistance to students and anyone else seeking information about the plants of the region. She was generous to a fault, leaving carefully curated cupboards of species she recognized as new for others to describe in the academic literature; maybe she preferred field work and the plants themselves to fighting with reviewers and the writing up of results. Although her influence was felt strongly amongst those who worked with and visited her in the herbarium to learn about the flora, she was not widely recognized in academic circles for her work. In 1989, however, she was awarded an honorary degree by the University of Cape Town, which she accepted, although with some reservations and what was typical self-deprecation:

> "... it does not seem to me that my work warrants the hon. degree. I hope Council were not misled into supposing that the discovery of undescribed species was a great achievement because it wasn't."

Of course, they weren't misled! She was much more than just discovering undescribed species.

Elsie Esterhuysen was a true field botanist, surviving for days in the field on a few vegetables and some powdered milk, or refusing to use air mattresses but instead bedding down on a pile of grasses or restios. She taught the many people she influenced much more than names of plants; she imparted to them her love of nature, of natural spaces and of the plants themselves; she showed them how to be "at home in nature". This constant and persistent dedication to the plants themselves is a rare and precious quality and can be hard to find in the pressured world of modern academia. That Elsie went back to collecting localities, where she had encountered novelties in flower, to collect fruits that might help in identification meant that her collections were not mere numbers, but were useful in, as she put it, "filling gaps". In the speech accompanying the presentation of her honorary degree she was likened to a botanical hunter-gatherer, "Like a hunter-gatherer she was in perfect harmony with nature, never being wasteful, never taking more than was needed at the time, never defacing or contaminating the environment." Such harmony with nature is something to which we all should aspire.

Esterhuysenia stokoei was originally described by Louisa Bolus as a *Mesembryanthemum*. This drawing held in the Bolus Herbarium in Cape Town, South Africa has delicately pencilled details of the fleshy leaves and fruit on the original.

Franklinia

BENJAMIN FRANKLIN

Family: Theaceae, the camellia family
Number of species: 1
Distribution: Georgia, USA, now extinct in the wild

Benjamin Franklin, that great polymath of the American founding fathers, might have preferred to have the plant that bears his name to be one that was useful, or common – but perhaps he also might have taken a certain pleasure in the one he got.

Born in Boston in 1706 as the penultimate of seventeen children of a merchant and candlemaker, Benjamin Franklin helped to shape a nation. One of the original band of "raw, undisciplined, cowardly men" – as the First Lord of the Admiralty in 1775 characterized the American rebels – who orchestrated the independence of the American Colonies from British rule, Franklin was the scientist-statesman among them. But this learning was not acquired in school or university; it was self-education at its finest. As a young man Franklin moved from Boston to Philadelphia, where he worked initially in the printing trade. He was sent to London as a diplomat several times, first in the service of King George III and later as a representative of the new American state. He established a postal service, wrote on the necessity for introduction of paper money, messed about with electricity, advocated the abolition of slavery, and believed passionately in the rights of man. Because he was an accomplished diplomat he also played key roles during the War of Independence itself, negotiating with the British generals – in particular he was one of the party of three who, on Staten Island in 1776, dashed the hopes of Lord Howe, commander of the British forces, by stating categorically there was no way back from "treading this step of independency." Franklin was one of the five men, along with Thomas Jefferson, John Adams, Robert Livingston and Roger Sherman, who drafted the Declaration of Independence, and later was instrumental in the establishment of the Bill of Rights, foundational documents for the independent American republic.

It was in Philadelphia that Franklin met John Bartram – a Quaker plantsmen whose knowledge of both horticulture and the plants of the region was unsurpassed. Both men were founding members of the American Philosophical Society in 1743, an organisation established to further scientific knowledge for use in the new colonies – and after 1776 in the new nation. John's letters to Franklin were addressed to

William Bartram's painting of *Franklinia alatamaha* was sent to botanists in England to convince them it was indeed new and different – he labelled it "a beautiful tree."

"Dear worthy Friend", and the two men regularly exchanged family news along with seeds and plants. John was a Pennsylvania farmer whose interest in the wild plants of the region led him to set aside part of his land for growing them, and to explore widely from Ontario to Florida. Not only did he cultivate new and interesting North American plants in his own garden – what could be said to be the first botanical garden in North America – but he also sent seeds to plantsmen in Europe.

John Bartram made at least fourteen long journeys to different parts of New England between 1734 and 1766, along with many shorter ones closer to home. This was not, however, without problems. Bartram had a farm to run in Pennsylvania, gardens full of interesting plants to keep up and he was active in the intellectual life of Philadelphia. Fortunately for Bartram he had a family of sons, each of whom helped in the family enterprise. One of the elder sons, also John, managed the farm, but the fifth son, William – affectionately known as Billy – was the one with whom John Sr. botanized.

William went with his father on an extended collecting trip to the Catskill Mountains (in southeastern New York state) when he was 14, and he never looked back – collecting and drawing natural diversity was his passion for the rest of his life, despite his father's attempts to have him take up some sort of trade or other more settled occupation. Despite William's difficulties in settling down to a 'proper' profession, he was an invaluable companion to John on the first of the collecting trips to the American South that yielded such botanical riches for his patrons in Europe.

During father and son's first trip to the South, they travelled up the Altamaha River from the Atlantic coast of Georgia, seeking rarities in seed to send to England. There they found "several very curious shrubs one bearing beautiful good fruite" – one of these was *Franklinia*. During the journey, lasting from July 1765 to April 1766, John kept a daily record and William sketched and drew the diversity they encountered. They travelled through the Carolinas and into Georgia and Florida, in part exploring for botanical riches, but also establishing British dominion over these southern areas. The region of St. John's River in eastern Florida was to be mapped and developed as an agricultural hub for British expansion; both Bartrams were present at the ceremony in 1765 that resulted in the Creek people ceding a huge part of northern Florida to the British. William

Benjamin Franklin (1706–1790) was not only a founding father of the United States of America and noted polymath, he was also a keen natural historian and gardener.

respected and valued the original inhabitants of the Americas; he later said "we ought to consider them as they are in reality, our Brethren and fellow citizens, and treat them as such…."

William's true loves were plants and drawing. That his drawings were good there is not a shadow of a doubt, even his father, in many ways his harshest critic, was pleased enough of them to send them to his patrons in Europe – saying "botany and drawing is his darling delight, am afraid he can't settle to business else". In 1772 William wrote to one of the Bartram's patrons, the Quaker physician John Fothergill of West Ham near London, suggesting a botanical collecting expedition to the southeastern regions of the Carolinas and Florida, with the aim of exploration and collection of botanical novelties – like father, like son.

William embarked south, alone this time, arriving Charleston, South Carolina in 1773. This journey would see him leave a colony and return to a new nation. He quickly set off for the south, and on the Altamaha River in Georgia, rediscovered the "curious shrub"

The type specimen of *Franklinia alatamaha*, grown in Philadelphia from seed collected in what is now Georgia, doesn't really do justice to the plant's beauty.

he and his father had seen almost ten years earlier. It appears that on this occasion he collected seeds, although the record is slightly hazy on this point – he later saw the plant again, in the same place a mere "30 miles from the Sea Coast" along the Altamaha River. William sent a drawing – said to be his best ever – of the plant to Fothergill, and from the seeds he collected, cultivated *Franklinia* back in Pennsylvania where it flowered for the first time in 1781 and eventually grew to 15 m (40 ft) tall. It was featured in the catalogue of plants from the Bartram's garden in 1783, as "Alatamaha, Undescript Shrub lately from Florida".

This trip took William to the banks of the Mississippi River, and deep into Florida – and by the time he returned to Pennsylvania in 1777, the colonies had achieved their independence from the British and were on the way to nationhood. It must have been extraordinary to emerge from several years of hardship and pleasure collecting to discover you were the citizen of a new nation. William's father John (who died shortly after his return) had been among those establishing the intellectual life of the colonies, centred around Philadelphia, in partnership with the giant intellects of Benjamin Franklin and Thomas Jefferson.

All plants of *Franklinia* now in cultivation are the descendants of the original trees grown from seed in the Bartram's garden in Philadelphia; it has never been found again in the wild.

The struggle to establish a distinctly American science and philosophy was central to the years following 1776, and Franklin was at the forefront of realising this national identity. William Bartram too played his part. His book about his journeys in the southeastern part of the United States with the fantastically long title of *Travels through North & South Carolina, Georgia, East & West Florida, the Cherokee County, the extensive territories of the Muscogulges, or Creek confederacy, also the country of the Choctaws; containing an account of the soil and productions of those regions, together with observations of the manners of the Indians*, published in 1791, was more than a laconic account of day to day sightings and journeys. It was one of the first natural history books that showed that the fauna and flora of the Americas was different from that of Europe, but also comparable if not surpassing it in diversity and distinctness. The language William used to describe landscapes, people and plants was intensely poetic, and reflected his own intense relationship with nature. In it he affirms the beauty of American nature, his own admiration of the native peoples' relationship and care for the environment. His prose influenced the Romantic poets such as WilliamWordsworth and Samuel Taylor Coleridge – his vision of American nature as sublime unity spoke to these men who rejected the Enlightenment view of nature as rational and ultimately understandable.

Despite originally thinking the lovely plant he found on the Alatamaha was a member of the genus *Gordonia*, William was convinced it was different and wanted to establish the name of the plant as *Franklinia alatamaha*, in honour of his father's friend Benjamin Franklin and the native American name for the river where it was found. But local botanists in the Americas did not have the reference collections to compare their new finds with, even though they did have access to some literature. William thus sent drawings to several European botanists attempting to persuade the botanical elite of Europe to accept his plant as a distinct genus. The name *Franklinia* had been published in 1785 by a Bartram cousin, Humphry Marshall – but the European botanists like Sir Joseph Banks insisted that it was merely another species of *Gordonia*. Deciding whether two genera are distinct depends upon comparison and judgement on the distribution of characters, the plant form, its leaves, differences in flowers and fruits. Banks obviously thought *Franklinia* too much like *Gordonia* to merit a different name, while Marshall and others recognized them as distinct.

Joel Fry, present day curator of Bartram's garden, has suggested that Banks's response to the distinctness of *Franklinia* was singularly lacking in botanical analysis, and that it may well have been intended to "enforce European supremacy in botanical nomenclature". It also might have been just too early and the success of the American War of Independence too raw to accept a name honouring one of the fathers of that struggle. Marshall and Bartram had a different scientific hypothesis as to the distinctness of *Franklinia* than did Banks and other British botanists, whether politics was involved or not. It's all about evidence.

Botanists using a broad suite of evidence from both morphology and DNA sequences today recognise *Franklinia* as distinct from its close relative *Gordonia*. Both are from the southeastern United States and are closely related to the eastern Asian genus *Schima*. These shared relationships between the southeastern United States and eastern Asia are common in flowering plants and are evidence themselves of our changing Earth. Both Bartrams found the plant, William cultivated it and gave us the name *Franklinia*, but he did even more than that. His seed collections, grown on in the garden in Pennsylvania in the newly established United States, are the sole source of the plant in cultivation. *Franklinia* was known only from a very small area of land in coastal Georgia and was last seen in the wild in the 19th century.

Benjamin Franklin helped re-establish the plant trade with Europe after independence, this time through France, where he served as ambassador. We don't know if he was pleased with *Franklinia* or whether discussions of the name for this unique American plant passed him by in his flurries of activity as ambassador to France and then governor of Pennsylvania in the late 1700s. But I think he would have been pleased to have a very special and very beautiful American tree named for him - especially one that had been discovered by a true friend and described by a staunch defender of emergence and independence of American science. Although the rose is the national flower of the United States, the story of *Franklinia* shows how plants themselves can help shape nations.

Gaga

LADY GAGA

Family: Pteridaceae, the brake fern family
Number of species: 18
Distribution: southwestern United States of America to Bolivia

Botanical taxonomists, those people who provide names for plants, are often portrayed as dull folk, huddled away in corners, surrounded with dusty Latin tomes and piles of dried plants on bits of paper. Lack of a sense of humour or any connection with popular culture seem to be prerequisites. Well, nothing could be further from the truth – naming plants can give people hours of fun and sometimes it allows them to demonstrate just how connected science and culture really are.

Naming plants for royalty was acceptable to Linnaeus – after all royals supported the science of botany by funding botanic gardens and other scientific establishments. But, he wrote, "generic names should not be misused in order to perpetuate the names of Saints and men [sic] distinguished in some other branch of learning…." All well and good for his day, but what of today, when we respect and admire people not for their position nor for their status, inherited or botanical, but for their deeds and accomplishments? For botanists are members of society too; they are not just concerned with plants to the exclusion of all else. Today we name plants for all manner of things, among them people we admire, whether they be botanists or royalty or not.

Taxonomists stuck to royalty, in a way, when they named the fern genus *Gaga* after the 'Queen of Pop', Lady Gaga. Here's their explanation:

> "The genus *Gaga* is named in honor of the American pop singer-songwriter-performer Lady Gaga, for her articulate and fervent defense of equality and individual expression in today's society. Because Lady Gaga speaks to the need for humanity to celebrate broad differences within its own species, we hereby provide her with a scientific namesake that characterizes the struggle to understand the intricate biology underlying cryptic patterns of diversity. Because public funding supports basic research, this naming honor allows us to acknowledge the confluence between science and public interests, and to make our findings more accessible and relevant to the diversity of individuals who fund our work. The name *Gaga* also echoes one of the molecular

Specimens of ferns, like this one of *Gaga marginata*, need to be arranged so future users can see all parts of the plant, including the undersides of the leaves.

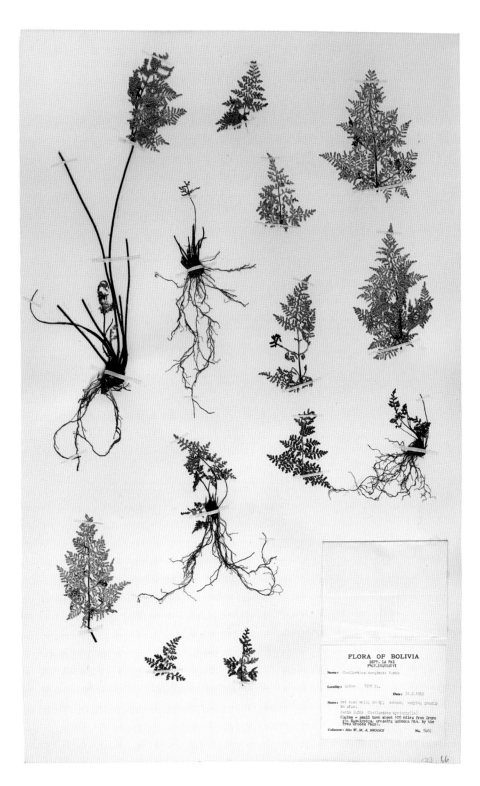

FLORA OF BOLIVIA
DEPT. LA PAZ
PROV. INQUISIVI

Name: Cheilanthes marginata Kunth

Locality: below 3500 ft.

Date: 31.3.1949

Notes: wet foot wall; shady; common; varying greatly
in size.

(with S.6A Cheilanthes myriophylla
(Quite a small town about 100 miles from Oruro
Río Huañiyapu, crossing Quimsa Mts. by the
Tres Cruces Pass).

Collector: Miss W. M. A. BROOKE   No. 5402

synapomorphies that characterizes the genus. At nucleotide positions 598–601 in the matK gene alignment, all *Gaga* species have "GAGA", a sequence pattern not seen at this site in any other cheilanthoid fern sampled [e.g. the closely related *Aspidotis densa* shows GAGG, and the type species of *Cheilanthes* (*C. micropteris*) has CAGG]."

By naming the new genus *Gaga*, these fern taxonomists commemorated not only their admiration for the person and her views, but the need to present their work to a wide diversity of people, not just other botanists. And lastly, they made a clever play on the base pair sequence found in the DNA of all species in their new genus. Would that the botanists of old had been so forthcoming about why they named genera for people!

Lady Gaga is known for her imaginative costumes and high-energy stage performances. Born Stefani Germanotta – Gaga chose her professional name from the song *Radio Ga-Ga*, written by Roger Taylor and famously performed by the legendary rock band Queen. Gaga's stellar career and versatile imagination have influenced many other singers and performers, and she has achieved stardom in film and on a world stage. Asked why they named a fern found mostly in Mexico after not a fellow botanist but a pop megastar not obviously connected with botany, the team explained that they "listen[ed] to [Gaga's] music while we do our research... We think that her second album, Born This Way (2011), is enormously empowering, especially for disenfranchised people and communities like LGBT, ethnic groups, women—and scientists who study odd ferns!" One of Gaga's memorable costumes also inspired the group – at the 2010 Grammy Awards her translucent green heart-shaped attire resembled a giant fern gametophyte – a key life stage in the fern reproductive cycle.

One of the outfits Lady Gaga (b. 1986) wore at the 2010 Grammy Award ceremony was green and heart-shaped, just like a fern gametophyte, but it wasn't the whole inspiration for the new genus name.

Ferns are not like flowering plants, which produce seeds, and their reproduction was a mystery for a long time. Early herbalists, firmly believing that all plants must have seeds, came to the logical conclusion that fern seeds must be invisible, after all no one had seen them. Legends abounded that the seed of ferns could only be collected on Midsummer's Eve, and once obtained it would render the bearer invisible, just like the 'seed' itself. Falstaff's co-conspirators in William Shakespeare's play *Henry IV* reckoned they were safe from detection in robbery – "we have the receipt of fern-seed, we walk invisible." Fanciful indeed. Eventually botanists began to examine the patches of 'dust' on the undersides of fern leaves; some reasoned these were the seeds, others that the 'dust' was like the pollen of flowering plants. It was the botanist John Lindley who, in the late 1700s, finally began to crack the puzzle. He sowed the 'dust' on bare soil and propagated fern plants that grew to maturity. This dust is not seed but spores; ferns are seedless plants and they reproduce by tiny spores, held in special structures on the leaf undersides. Once released the spores develop into a free-living structure called a gametophyte, so called because this is the life cycle stage that produces the gametes, eggs and sperm. The plant we think of as a fern is the sporophyte, it produces the spores and is asexual. Ferns and their relatives are unique among plants in that they have a free-living gametophyte, an independent plant not connected to the sporophyte in any way. Fern gametophytes are small, green and photosynthetic, and often filamentous or heart-shaped – like Gaga's costume at the Grammys.

The spores of *Gaga* species are held under the rolled-over leaf margins, protecting them from desiccation until just the right time for their release.

In many botany textbooks the fern life cycle is depicted as a closed loop in which the sporophyte (fern body) produces a spore that then germinates into a gametophyte that produces both eggs and sperm that meet and join to form the zygote. This, in turn, gives rise to the sporophyte, and the cycle begins again. The gametophyte and sporophyte generations alternate – alternation of generations is one of the concepts fundamental to the teaching of botany. The alternation is between the gametophyte with one set of chromosomes and the sporophyte with two. First the process of meiosis gives rise to the spores that develop into the gametophyte, then the gametophyte develops structures called archegonia that produce eggs and antheridia that produce motile sperm (that swim in water with flagellae), then egg and sperm join via fertilization, resulting in the growth of the sporophyte, the plant body we recognise as a fern. Because gametophytes can produce both eggs and sperm and potentially then self-fertilize, ferns were thought to be examples of extreme inbreeding, leading to a lack of genetic variation and thus limited potential for evolution. Some attributed the lower species numbers of ferns as compared to flowering plants to this presumed inbreeding in bisexual gametophytes. But as usual with living organisms, the story is more complicated than that.

As free-living organisms, gametophytes are amazing. They can exist as populations for years, never producing sporophytes. In the United Kingdom and Ireland, the Killarney fern, *Trichomanes speciosum*, is known mostly from populations of gametophytes; sporophytes are very rare indeed. Living at the edge of its range, the gametophyte populations of this fern persist, producing sporophytes only rarely and perhaps in response to milder weather regimes. Gametophytes are resilient, withstanding extremes of heat and cold, reproducing vegetatively until the time is right. Far from being automatic self-fertilizers, fern gametophytes exhibit a wide range of breeding systems and sex expression. Early-maturing populations of gametophytes can produce eggs and then secrete a hormone called antheridogen into the substrate that induces later-developing gametophytes to produce only sperm. In systems like these there is obviously cross-fertilization, and that process leads to an increase in genetic variation. This can happen between gametophytes from spores from the same sporophyte (selfing), or between gametophytes derived from spores of different sporophytes (outcrossing) – the possibilities are endless. Outcrossing is predominant in breeding systems that have been studied in ferns; they are most emphatically not extreme inbreeders.

Ferns can also produce spores without meiosis, leading to gametophytes with two sets of chromosomes that can produce sporophytes directly from the gametophyte body, no eggs and sperm needed. This is asexual reproduction and is particularly common in ferns that grow in dry habitats, like *Gaga*. This mode of reproduction is advantageous in arid areas because it eliminates dependence on water needed by swimming sperm and shortens the time for sporophyte production. Of the 19 species of *Gaga*, about half are thought to reproduce in this way. These species – among them *Gaga germanotta*, the species named in honour of Gaga's family – have odd numbers of chromosomes, making regular chromosome pairing impossible; they are entirely asexual.

All this variation in breeding system and sex expression means that ferns are incredibly versatile when it comes to establishing new populations in emerging habitats. Ferns are often the first plants to appear on lava flows following volcanic eruptions – a population can establish from a single spore, persisting until others arrive to inject genetic variation. So, the fern life cycle "should not be seen as a 'weak link' or vestige of early land plants but rather as an intricate adaptation that has supported the long and continued success of ferns…."

The chelianthoid ferns, of which *Gaga* is a member, are

The leaves – or fronds as they are usually called in ferns – of cheilanthoid ferns have rolled margins and are often hairy, scaly or floury on the undersides.

characteristic of dry forests and deserts; they are small plants with remarkable adaptations to life in the face of water shortages. In times of extreme water stress the leaves curl up, minimizing the surface exposed to the sun, uncurling when water is plentiful again. Leaf edges curl over to protect the developing spores, looking like they have been crimped or very slightly folded over, very different from the exposed patches of spores seen in some other ferns. The leaves and stems are often covered with hair or scales that absorb water directly, getting it straight into the plant body without waiting for it to be translocated from the roots. Chloroplasts - the photosynthetic factories in these leaves - are protected from UV damage that can

Gaga germanotta was named by botanists in honour of the Germanotta family and in recognition of their tireless work towards a kinder more inclusive society.

arise from growing in high light environments by pigments or pigmented structures. Sometimes these ferns lose their leaves entirely, the plants remaining as brown stalks! These are not your typical ferns from dark, wet forest understories; these ferns have adapted to a totally different way of life. These specialized adaptations to life in dry environments coupled with widespread asexual reproduction has meant chelianthoid ferns, and especially the genus *Cheilanthes*, have until recently been seen as "the most contentious group of ferns with respect to a practical and natural classification."

DNA sequence data has helped with the recognition of monophyletic groups – those groups that contain all the descendants of a common ancestor – within *Cheilanthes*; *Gaga* is one of these monophyletic groups. Most of the species of *Gaga* were previously classified as *Cheilanthes*, a large genus whose classification is still a work in progress. There is much left to learn - about fern biology, about classification of taxonomically difficult groups like the cheilanthoid ferns, and about fern species.

In recognising *Gaga*, two new species were also distinguished – *Gaga germanotta* and *Gaga monstraparva* – the first honouring the Germanotta family and specifically their Born This Way Foundation, established to empower young people to create a kind and accepting world, and the second an allusion to Gaga's legions of fans, the 'little monsters', whose claw-hand paws up signal somewhat resembles an unfurling fern frond. But what does the Queen of Pop think of all this? When asked in a 2014 Reddit Q&A session how she felt about having a fern genus named after her, Lady Gaga replied "It's pretty cool, especially since it's [an] asexual fern, there are 19 species contained within the genus. All sexless, judgeless. How cool. How I wish to be."

Hernandia

FRANCISCO HERNÁNDEZ

Family: Hernandiaceae, the lantern-tree family
Number of species: *c.* 30
Distribution: circumtropical

The first time I saw plants of the genus *Hernandia* in the Central American tropics I fell in love with their wonderful leaves. Large, green and slightly rubbery, these leaves are what botanists call peltate; the petiole, or leaf stalk, seemingly arises from the middle of the leaf blade making the whole thing look like an umbrella. These fantastic leaves are aromatic and full of essential oils – some species have been used in the manufacture of perfumes. *Hernandia*, and its eponymous family, are members of one of the early branching lineages of the flowering plant evolutionary tree and are related to the laurels; the high content of essential oils is one thing the two families have in common. The pollen (male gamete) bearing organs – the anthers – in both the laurel family and *Hernandia* open by tiny little flaps, looking for all the world like a trap door, rather than by elongate slits like most other anthers; this too is a shared derived characteristic hinting at evolutionary relationships.

In plants, most flowers function as both male and female, containing both stamens that hold pollen and an ovary that holds the ovules that will become seeds. But separation of sexual function is widespread amongst flowering plants and comes in many variations; sex expression in flowering plants is not a simple system. Plants of *Hernandia* can be either monoecious – with the flowers of both functions borne on one tree – or dioecious – with the flowers of different functions on separate trees. To see it from a plant's perspective – we human beings are dioecious – our sexual functions are done by different individual organisms.

The 'fruits' of *Hernandia* are red or yellowish cream and are probably dispersed by animals, although some species apparently drop the fruits into water, after which they float downstream or across oceans, perhaps accounting for the wide distribution of the genus. The species *Hernandia nymphaeifolia* is found on beaches all over the Pacific, where its light wood is used in canoes, and its seeds are used as jewellery. One of the common names of this species is the lantern tree – probably from the shape of the fruit that looks a bit like a light fixture, with a fleshy lampshade derived from the tiny leaf-like structures underneath the flower covering the hard black lightbulb-like true fruit.

Sydney Parkinson was the artist aboard HMS *Endeavour*, but died before making it home; his painting of *Hernandia nymphaeifolia* from Tahiti captures the odd leaves and fruits of this plant perfectly.

Hernandia ovigera.

"Otaheite"

Sydney Parkinson pinx 1769.

Someone once said that Hernández was paid a lot, but produced nothing, just like the fruits of *Hernandia*, plump and juicy looking but only hollow shells – nothing could be further from the truth.

The Swedish botanist Carl Linnaeus is credited with coining the name *Hernandia* in his great work *Species Plantarum*, in which he 'invented' the naming system we still use today for the scientific names of all plants and animals. Botanists use the 1753 publication date of *Species Plantarum* as the starting point for all scientific naming, but names were being given to plants long before that, both by peoples who used them in their native lands and by European explorers and botanists who were struggling to understand the world around them. Linneaus 'borrowed' the name *Hernandia* from the French botanist Charles Plumier, who had published a book describing new American genera of plants 50 years earlier. Plumier dedicated this new plant to Francisco Hernández, the Spanish physician sent by King Philip II to carefully observe "plants, liquors and other physicians" in the territory of 'New Spain' – today's Mexico.

In his own time the Spanish physician Francisco Hernández was known as 'the Third Pliny', illustrative of his desire to document and disseminate information about the natural history of everything, as did the Roman physician Pliny in the 1st century of the modern era. He was trained as a medical doctor during the early part of the 16th century, probably at the University of Alcalá de Henares to the north of Madrid, a centre for humanist thought in the Spain of the time. In the 16th century the study of medicine was still strongly underpinned by the study of botany, as medicines were almost completely herbal in origin; thus, to be a physician, one had to have a good knowledge of plants and natural history as well. Hernández was certainly no exception and in his later life he would characterize himself as a 'doctor and historian', signalling his desire to understand the total history of the world.

To rise in the medical profession in 16th century Spain it was necessary to be near the court, go to sea or go into the church; Hernández clearly chose the court and in 1567 was appointed royal physician to Philip II. The image of Philip II, who married Mary Tudor whose attempts to reintroduce Catholicism into the English monarchy were at odds with her half-sister Elizabeth I, is often one of a dour man, dressed in black and presiding over a Spain in the grips of the Inquisition and religious mania. But he was also a well-educated man whose interests were broad and included the 'New World', whose peoples, fauna, flora and geography were so poorly known to the Europeans. He had the monumental library at the Escorial palace constructed, together with research laboratories into the efficacy of medicinal remedies – all this associated with his own royal apartments.

In 1570, only three years after his appointment as a royal doctor and at the age of 55, Hernández was tasked by Philip with the role of 'protomédico' of the Indies – the chief medical officer for the lands in the Americas conquered by Spain. In his marching orders it is clear that the roles of physician and explorer were intimately intertwined. The orders specified that Hernández should only go to those lands conquered by Spain, because there were "more plants, herbs and medicinal seeds are to be found there than elsewhere", that he should consult widely with local doctors, including native peoples, to acquire as much medical knowledge as possible to be of use back in Spain, and that he should write down descriptions of plants and their habitats. And after he finished with 'New Spain' he should go on to Peru and do the same thing there. Pretty tall order for a land only half-understood by Europeans.

So Francisco Hernández set off for 'New Spain' with a giant task. He was obstructed from the start by the local

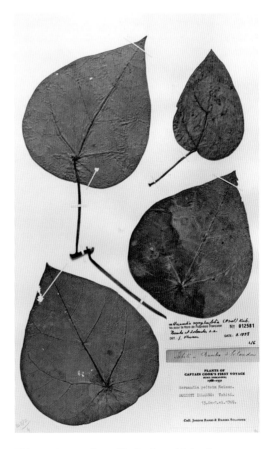

This specimen, collected by Banks and Solander on Tahiti in 1771, clearly shows the umbrella-shaped leaves of *Hernandia nymphaeifolia* that so capture the imagination.

Spanish bureaucracy, who had little time for the wishes of a monarch so far away and, presumably, for an intellectual who might show them up as not doing their jobs properly. He was even called up by the Inquisition – which operated everywhere in the Spanish dominions – to give evidence against a fellow physician, perhaps a veiled warning. His job as Philip's 'protomédico' largely blocked, Hernández threw himself into his other task, documenting the plant life of these extraordinary lands.

This he did through his own observations and travels in the area around Mexico City and further afield, and by talking to local peoples about the uses and properties of the plants he saw and described. His methods appear to have been similar to those used by others at the time, a formulaic set of questions that captured information in a consistent way – What is it? What is it called? How is it used? Where does it grow? He also often commented on the possibilities for cultivation in Spain – how could not only the knowledge but the actual plants themselves be brought back for royal benefit.

Hernández spent about six years exploring Mexico, travelling to major hospitals in the area around Mexico City and also further afield as far as Michoacán and Colima to the west and Oaxaca to the southwest. His travels have been reconstructed not through diaries or travel journals, but through the descriptions of the plants he saw and the people he consulted. He wrote thousands of descriptions, and because he was accompanied on these journeys by native Mexican artists, each plant was illustrated. The scale of this endeavour was huge – imagine documenting the diversity of one of the world's biodiversity hotspots from a completely blank page.

Hernández wrote annual letters to Philip, documenting the state of his work as it progressed. In 1571 "the natural history of the Indies is proceeding with all proper care and diligence… more than eight hundred new plants, never seen before in these parts have been depicted". But by 1576 he was getting fed up – "Sacred Catholic Royal Majesty. I have delivered to the royal officials, so they can send them to your Majesty, with the fleet that is now ready to leave New Spain, sixteen large volumes of the natural history of this land… The work, that has cost me my health and my life, promised much benefit and is now beginning to show some…" He eventually returned to Spain in 1577 in extremely poor health, with copies of his manuscripts in Latin, Spanish and intriguingly with the intention of translating the whole into Nahuatl. He never saw the publication of his work in its entirety.

Francisco Hernández (1514–1587) never published his work himself, but we are lucky that extracts were copied since the top copy was destroyed in the great 1671 fire in the royal collections of the Escorial.

Hernández ordered his work not by medical usefulness or by botanical affinity as understood at the time, but by the local Nahuatl names for the plants he described. He demonstrated respect for Nahua traditions and knowledge, writing that "I show how virtuous these people were even when they were idolaters and cannibals, and how much care they took to educate their children, and how great was the power of their sayings". Notwithstanding his condemnation of the peoples' state before the advent of Christianity, he opines that virtues in 'New Spain' have arisen independently of European values. Hernández clearly valued the local people with whom he worked as individuals – he left money in his will to three of the artists whose paintings contributed so much to his work.

Hernández tinkered with his work, copying and re-copying, refining and editing; it was to another that, in 1580, Philip II gave the task of making a selection of the descriptions for publication. This selection was the basis for the dissemination of Hernández's work for centuries and the basis for European natural historians' knowledge of the biodiversity of Mexico at the time. Hernández's detailed descriptions introduced to the European world plants such as cacao, maize, chilis, tomatoes and vanilla, not just as commodities but as plants whose use was woven into the culture and health of the peoples where they grew naturally – he "decided to write only about the things that are familiar to the New World, but still not sufficiently known in ours."

His writings were unique in that they were based on Nahuatl knowledge and culture, and they were published in their entirety in Mexico only in the latter part of the 20th century. It is fitting that one of the partial copies of Hernández's work led botanists of the 18th century to name such an interesting plant as *Hernandia* in his honour.

In Jamaica, the endemic *Hernandia catalpifolia* is the larval food plant for the endangered Homerus swallowtail, *Papilio homerus*, the largest butterfly in the western hemisphere.

Hookeria

WILLIAM JACKSON HOOKER

Family: Hookeriaceae, the shining-moss family
Number of species: 10
Distribution: worldwide, except tropical Asia

The great early 20th century artist Georgia O'Keefe once said "When you take a flower in your hand and really look at it, it's your world for the moment." I couldn't agree more, plants are infinitely absorbing. But mosses take us to a whole other level. Often passing almost unnoticed as the green fuzz on the tops of walls or on rocks in the forest, mosses are worlds in themselves. Look at the green atop a garden wall closely and a tiny forest appears; look under a microscope and tiny animals like tardigrades, or water bears, gambol about with unicellular organisms of many sorts. It's biodiversity as you have never thought of it before.

Mosses are plants that reproduce by spores, and together with the ferns and lycophytes are called cryptogams – from the Greek meaning hidden reproduction. This contrasts with the phanerogams or seed-bearing plants, whose reproductive organs are more obviously out in the open, most obviously of all in the flowers of the flowering plants. Mosses and their relatives the liverworts and hornworts – collectively called bryophytes – lack a vascular system; they do not circulate water or nutrients around the plant body via xylem and phloem, but instead absorb nutrients more directly through their tiny leaf-like microphylls and root-like rhizoids.

Bryophytes are often characterized as 'primitive' plants because they lack the innovations like a lignified vascular system or flowers that have made other lineages of plants so successful in terrestrial environments. Evolutionarily, however, this is the wrong way to describe them – the lineage leading to today's living bryophytes has evolved and changed over the same time period and for just as long as that leading to vascular plants. It is more accurate to describe them as sister lineages – they split from a common ancestor a long time ago and have followed different trajectories ever since. And it was a long time ago that these two plant lineages shared a common ancestor – estimates vary, but it is clear that plants like bryophytes have been with us since at least 400 million years ago, when life was first emerging onto land. Bryophyte-like organisms have been part of terrestrial life on Earth for a very long time; they saw the dinosaurs come and go and carried on evolving through several mass extinctions.

Mosses might all look the same from a distance, but their tiny leaves, capsules
and rhizoids have a wealth of difference. *Hookeria,* with its thin rounded leaves,
is in the centre of this illustration.

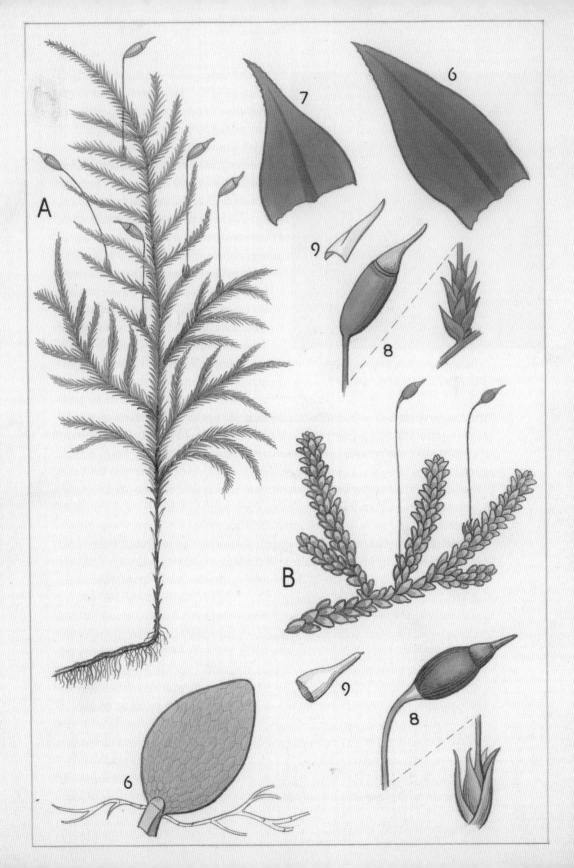

Hooker take the exciting little moss to Mr. Dawson Turner of Yarmouth, a banker who, like many of the monied classes in England at the time was a keen bibliophile and natural historian. Turner set the young Hooker to work drawing for his book on British algae – Hooker was not only a keen naturalist, but also an accomplished artist; a fellow botanist remarked "I can hardly say what I admire more in his works, his pencil or his pen." Both Turner and Smith saw great potential in Hooker – Smith was concerned he would abandon mosses and bryophytes for insects or birds, and Turner was keen to have his assistance in his own publications. Smith gave Hooker the run of the Linnaean collections, and between him and Turner introduced Hooker to the great and the good of the natural history world of the day.

A couple of years after Hooker's collection of *Buxbaumia*, Smith dedicated a genus of moss to his young friend (FLS stands for Fellow of the Linnean Society, to which Hooker had been elected as probably the youngest member ever at the time):

> "I have great pleasure in dedicating this genus to my young friend Mr. William Jackson Hooker of Norwich, F.L.S., a most assiduous and intelligent botanist, already well known by his interesting discovery of *Buxbaumia aphylla* as well as by his scientific drawings of Fuci for Mr. Turner's work; ..."

Hooker himself illustrated Smith's work and later published his own assessment of the genus named for him, removing several of the species that had been included by Smith as not truly belonging to the genus – Hooker on *Hookeria*!

Hooker had a great desire to travel the world and he got as far as Scotland and the continent. When Sir Joseph Banks presented him with an opportunity to go to Iceland he leapt at the chance; the trip to Iceland was a real adventure, topped off by the ship in which he was returning catching fire and being destroyed, resulting in the complete loss of all his collections and papers. So, back he went to his bryophytes. By this time, he had fallen in love with the eldest of Turner's daughters, so when Banks offered another trip, this time to Java, two families, his own and the Turners, worked hard to dissuade him. Banks was cross when he decided not to go but supported Hooker to the hilt in his application for the vacant professorship

William Jackson Hooker (1785–1865) studied the liverwort genus *Jungermannia*, producing a series of tomes on these tiny plants that were among his first botanical publications.

of Botany at Glasgow University. With his growing family Hooker moved to Glasgow, to give lectures to students of medicine. He was an astounding success – his lectures became popular for students and local naturalists alike. He used his artistic skills to demonstrate plant morphology in class and papered the walls with huge drawings of different plant families and genera. He also wrote a *Flora Scotica* for his students to use in the field. This was botany as it had never been taught; open, accessible and there for everyone. And his students certainly appreciated him – his first-year students presented him with a vasculum, the plant-collecting

The shining hookeria, *Hookeria lucens,* forms large mats in moist places in wet forests in temperate zones worldwide, like this patch from the rainforest in Wales.

case used in the days before plastic bags, engraved with fronds of *Hookeria lucens*.

Meanwhile, he also amassed a world-class herbarium and botanic garden that became a mecca for botanists from all over the world. He carried on his bryophyte work and published a guide to the mosses of Britain that has almost a manifesto of botany for the people in its introduction:

> "The English language has been preferred for this work, because we know
> many naturalists who pursue the study of this pleasing branch of natural
> history with the most unwearied industry, who are nevertheless in a
> situation of life which has precluded them from acquiring the knowledge
> of any but their native tongue; …"

His zeal in bringing botany to everyone was also reflected in the popular botanical magazines he was involved with, often contributing not only text but also illustrations, often at his own expense. His commitment to his students and to the gardens and herbarium in Glasgow was absolute, but he longed to return to London – his eye had long been on the Royal Garden at Kew, the emporium for the botanical riches of British exploration where, in 1843, he became director and began the next phase of his botanical career. His rich personal herbarium came south with him, along with his collaborative and inclusive spirit. Under his leadership the gardens opened for the public to enjoy the rich collections with a guidebook written by him. Hooker was a true champion of botany for everyone.

Juanulloa

JORGE JUAN AND ANTONIO DE ULLOA

Family: Solanaceae, the nightshade family
Number of species: 10–15
Distribution: southern Mexico to South America

Plants of *Juanulloa* are unlikely nightshades. Rather than being herbs or shrubs, as are most members of the family, like the deadly nightshade (*Atropa belladonna*) or tobacco (*Nicotiana tabacum*), species of *Juanulloa* are epiphytes, growing high in the tree canopy of tropical rainforests in the Americas. Epiphytes are plants that grow on other plants, from the Greek 'epi' meaning upon, and 'phyton' meaning plant. Species of *Juanulloa* are shrubs with thick leathery leaves whose beautiful flowers are striking, but not collected very often as they are hard to access.

Because few collections are made of these hard-to-reach epiphytes, their taxonomy and relationships are only now being clarified using DNA sequencing techniques. *Juanulloa* is a member of a lineage in the Solanaceae that are all epiphytic or hemiepiphytic, and whose flower shapes and sizes are hugely variable. Some have large, bell-shaped flowers, others tiny green tubes, while still others are held in tightly packed groups resembling footballs. *Juanulloa*, as traditionally defined, has waxy tubular orange flowers, probably pollinated by hummingbirds, whose diversity in the tropical forests where these plants occur is extraordinary. The floral tube is often slightly curved, perhaps fitting the bills of particular species of hummingbirds as is the case in other hummingbird-pollinated plants. The thick corolla tube is enclosed in an equally thick and waxy calyx – both these structures may serve to protect the nectar at the bottom of the tube – the reward for the hummingbird – from nectar robbers who cut through the base of the flower to gain 'illegal' access to the sugary solution.

Recent studies of the lineage to which *Juanulloa* belongs have revealed exciting new subgroupings and have overturned the traditional concepts of genera in these wonderful plants. Calling something a genus is in a way a matter of how a botanist divides up the phylogenetic tree of relationships – these trees are depicted as dichotomously branching diagrams and deciding which level of branching to call a genus can be tricky. Because taxonomy is not only a pure science used by taxonomists to study evolutionary patterns, but is also a tool for understanding diversity, managing conservation areas and many other investigations into the natural world,

The large nectaries at the base of these *Juanulloa parasitica* flowers secrete a sugary reward for the hummingbirds that visit and pollinate these epiphytes high in the canopy.

4118.

W. Fitch del.t   Pub by S. Curtis Glazenwood Essex Oct.r 1. 1844   Swan Sc

Juanulloa parasitica was so named by Ruiz and Pavón because its epiphytic habit made them think it was parasitizing the trees it grew on, rather than just using them as support.

how we name things is a balance between perfection and practicality. These decisions are sometimes not easy!

The genus *Juanulloa* was traditionally defined as being all the species of epiphytes in this larger lineage with tubular orange flowers; other species with smaller greenish flowers were defined as different genera. DNA sequencing work has shown however that some traditional *Juanulloa* species with the typical orange tubular flowers are actually more closely related to species with tiny greenish white flowers that were named as the genus *Hawkesiophyton*. So here we have a classic taxonomist's conundrum – to 'split' (leave as two genera with orange flowers in both) or 'lump' (merge into a single genus with both types of flowers). Botanists working with these plants will decide, publish their results and argue one way or another, and the paper will be reviewed by peers. The decisions then go into the literature to be tested by future botanists – every change we make is the sign that we are gaining knowledge. But whether it is one genus or two, *Juanulloa* will remain a name that we use; it was described by the Spanish botanists Hipolito Ruiz and José Pavón long before *Hawkesiophyton* was described, so as the oldest name it has what we call priority. If the two names are merged, *Juanulloa* is the one that will be used.

Ruiz and Pavón were botanists sent by King Charles III of Spain to document the riches of Peru and Chile in what was called the *Expedición Botánica al Virreinato del Perú* (Botanical Expedition to the Viceroyalty of Peru). Their stay of 12 years in Peru and Chile resulted in an explosion of knowledge about the plant diversity of Peru, today one of 17 'megadiverse' countries with exceedingly high biodiversity and traditional knowledge. As dedicated botanists they collected and described for science many new genera, among them *Juanulloa*, that they named for two illustrious Spanish naval men of the Enlightenment who had preceded them in the region some 50 years earlier. They dedicated the genus to Jorge Juan and Antonio de Ulloa in both Spanish and Latin, unusual at the time when Latin was the language of scholarship:

"Género dedicado a D. Jorge Juan y D Antonio de Ulloa, que acompañados de los Señores La Condamine, Joseph de Jussieu y de otros insignes Matematicos y Botanicos recorrieron el Perú con el fin de medir un grado del Equador para determiner la figura de la tierra, y publicaron varias noticias de Plantas de America de su viage impresa en Madrid año de 1748." [Genus dedicated to Don Jorge Juan and Don Antonio de Ulloa, who accompanying the scientists Condamine, Joseph de Jussieu and other luminaries of mathematics and botany, travelled through Peru with the aim to measure one degree of latitude at the equator in order to determine the shape of the earth, and who published in Madrid in 1748 various descriptions of American plants.]

Juan and Ulloa were members of the French-led geodesic expedition whose aim was to test Isaac Newton's then-controversial hypothesis that the Earth bulged at the equator and flattened at the poles. Newton's idea was much contested in scientific circles and in the early 1700s Louis XV of France decided to send two expeditions to settle the matter, one to the pole and one to the equator. The easiest place to reach on the equator at the time was the Spanish-colonized western part of South America –

in the area around Quito, in what is today Ecuador. But access to Spanish-held lands was not easy, the rulers of Spain jealously guarded not only the riches, but knowledge about the territories they had conquered in the previous century. Philip V of Spain granted permission for the French team to enter the region, but on the condition that they had with them two Spanish scientists as assistants to protect his royal interests. Jorge Juan and Antonio de Ulloa, both naval officers from Cadíz, were appointed by the royal court. Ulloa was only 19 years old, and Juan only 22 – extraordinarily young to be given such a charge, but both were well-versed in science, navigation, engineering and, as became useful later, military matters.

Ulloa in particular was a stickler for protocol and the right way to do things – he fell out with bureaucrats throughout his time in Peru. But the French team respected the knowledge and assistance of the two young men in their scientific quest. Juan

Nested within *Juanulloa* is the genus *Hawkesiophyton* (named for the botanist J.G. Hawkes) with small green flowers pollinated by bees – taxonomists are still deciding whether to 'split' or to 'lump'!

and Ulloa worked with the geodesic expedition until the Frenchmen left in 1739 – some by the traditional route north via what is now Colombia and La Condamine by boat down the Amazon. The young Spaniards stayed for 11 years in South America, fulfilling their royal instructions to join the French team, take part in all their studies, record the results carefully, make plans of all the cities, harbours and fortifications wherever they went, gather information about the soil, plants, industry and people of the region, including the 'uncivilized' indigenous peoples, and make observations that would be useful for navigators in the future. Tall orders for two young men in their early twenties.

Far from being cosseted and remaining with the Spanish bureaucrats and functionaries in the cities, Juan and Ulloa travelled widely into all sorts of different types of villages and situations – really recording what life in the Spanish colonies was like. The book they wrote upon their return, *Relación histórica del viage a la America Meridional* [Voyage to South America], was widely read despite being five weighty volumes. It was a real look into the reality of life, natural history and geography of the Spanish colonies; this is the book to which Ruiz and Pavón referred in their dedication of *Juanulloa* to Jorge Juan and Antonio de Ulloa.

But this account of their travels is not the only thing that they wrote – well actually, Ulloa wrote, as he did most of the writing, although Juan's name was listed before his on the title page because he held the superior naval rank. They also produced an account of the iniquities of the government of the colonies, for the eyes of the king and his ministers only – the *Discourse and political reflections on the kingdoms of Peru*. In it, no punches were pulled; widespread corruption and smuggling, mistreatment of local people by government officials and clergy alike and the general poor state of morality and service to the king were documented in excruciating detail. The *Discourse* documented mismanagement and appalling treatment of local peoples, but without naming names. In fact, these activities were said to be a "universal malady", undertaken by many sent to the colonies as administrators, functionaries and priests in order to enrich themselves and their families. In short, a get-rich-quick-scheme. The rigid social stratification of colonial society led to intense in-fighting between European immigrants and locally born people of European descent – "no other sure means to extinguish the flames of factionalism – an evil embedded in Peru almost since the time of the conquest – than by dissipating the power of all Europeans migrating to the Indies." Not words to make one popular amongst those who were in charge on the ground in Peru.

But the strongest words and condemnation was reserved for the treatment of the native peoples of the region. The punishing taxation and tribute systems, forced purchase of shoddy goods at what were known as 'repartimientos', land-grabbing under the flimsiest of pretexts and general exploitation of the local populace by administrators and clergy alike were recorded, based on fact and observation "in their villages we had sufficient opportunity to be a witness to both their complaints and well-founded outcries concerning the excesses and injustices inflicted on them."

Previous assessors had either been bribed to stay quiet, or just didn't care. Ulloa raged against the system, which he considered unjust in the extreme. His solutions ranged from extending the terms of colonial administrators, so they didn't see appointment as a way to get rich quick, to limiting the power of all Europeans, to making the sons of local leaders in charge of taxation matters. His solution to the appalling behaviour of the clergy was to put the Jesuits in charge.

The *Discourse* remained a secret document until 1826, when an unauthorized copy was made by David Barry, a British citizen who had ties with South America. Coming just after the independence struggles in South America, his transcription – entitled *Noticas secretas de America* – emphasized the "cruel oppression" and "scandalous abuses" and served to stoke anti-Spanish feeling, something that would have horrified Jorge Juan and Antonio de Ulloa. They remained loyal and steadfast servants of the kings of Spain until their deaths. Ulloa in particular was sent on various extremely difficult missions – as governor of the Peruvian region of Huancavelica, right into the web of corruption involving silver mines and, just as he was returning to Spain from that difficult post, to Louisiana as governor of the lands newly acquired from the French. His moral, almost puritanical, stance on corruption and mistreatment meant he was not well-liked in either place. Juan had a distinguished naval career and both men were elected to various European scientific societies.

In describing widespread corruption and mistreatment of local peoples, Juan and Ulloa intended their observations to spur the authorities in Spain to improve the governance of 'their' colonies for the good of all. It is not clear what became of the advice; the expulsion of the Jesuits from the Americas in 1767 clearly showed the advice was not necessarily welcome. By the time Ruiz and Pavón followed in Juan and Ulloa's footsteps to Peru in 1777 nothing else had really changed. The impact of Juan and Ulloa's travels and observations was strong in their own time, but they are largely forgotten today. They will, however, always live on together in the genus *Juanulloa*.

Antonio de Ulloa (1716–1795) was severely critical of the corruption rife in local bureaucracy of the Viceroyalty of Peru, particularly the treatment of local people; this did not make him a popular member of that society.

Lewisia

MERIWETHER LEWIS

Family: Montiaceae, the spring beauty family
Number of species: 16–18
Distribution: western North America to northern Mexico

Had Meriwether Lewis and William Clark and their 30-person strong 'Corps of Discovery' turned back from crossing the Rocky Mountains on the Lolo Trail on the border of what are now the states of Idaho and Montana in mid-June of 1806, we might never have had a plant genus named *Lewisia*. After spending a miserable rainy winter on the Oregon coast, Lewis was anxious to return to the east, laden with knowledge about the lands west of the Mississippi River. The Nimiipuu people (also known as Nez Percé) with whom Lewis and Clark and their team were staying strongly advised against starting out across the mountains. The route was their traditional path to the buffalo hunting grounds of the plains, but they said it was too early; there would still be too much snow and little grass for the horses. But Lewis knew better – he thought anything the native peoples could do he and his well-honed team could do too, so in the absence of guides and against all good advice, they set off, not without some trepidation though; on the evening before they departed Clark wrote "Even now I Shudder with the expectation with great difficuelties in passing these Mountains."

And difficult it was. Imagine a group of more than 30 people and horses attempting to cross a complex chain of high mountains, first in freezing rain, then snow, then encountering snowpack of 3–4 m (7–11 ft), no grass for the horses, with dramatic elevation gains and slippery trails. After about a week Lewis, consulting with Clark, came to a decision, "If we proceeded and became bewildered in these mountains the certainty was that we should lose all our horses and consequently our baggage instruments and perhaps papers and thus eminently wrisk the loss of the discoveries which we had already made if we should be so fortunate as to escape with our lives. ….. We conceived it madness in the stage of the expedition to proceed without a guide."

So, they turned back, waited for a week or so and proceeded to cross the Rocky Mountains with several teenage Nimiipuu boys as guides in six days, only one of which did not have grass sufficient for the horses. The previous year, they had taken eleven days to cross; their guide, the Shoshone (or Newe) man Pikee Queenah (Swooping Eagle) who was called 'Toby' by the expedition members, had got lost along the trail

The flowers of *Lewisia rediviva* are spectacular, but it is the starchy roots that both provided much-prized carbohydrates for indigenous people and give the plant its common name of bitterroot.

Montana

Mary E. Eaton

amongst fallen timber and blocked ravines and all nearly perished before reaching the plains of central Idaho. In contrast, in 1806 they reached the broad plain of what they called 'Traveler's Rest' in much better shape. There Lewis collected plants, among them *Lewisia* – a small tap-rooted herb that grows in gravelly soils – perhaps along the Lolo Creek that runs near the place the expedition made their camp, in a grassy plain that had long been a meeting place for indigenous peoples of the region.

The genesis of what is today known as the Lewis and Clark expedition was Thomas Jefferson's purchase, for US$15 million, of 'Louisiana' from Napoleon Bonaparte in 1803. Jefferson, the third President of the United States of America, was fascinated by the lands to the west and their potential for increasing the size and power of the nation. Napoleon, on the other hand, was pleased to get rid of lands he could not govern and to re-establish an alliance with the United States against the British. But just what had the United States government bought? The area to the west of the Mississippi River was little known to Europeans except for the solitary fur trappers who regularly voyaged up the rivers for the British and French-owned consortia.

Jefferson had the idea that an all-water passage could be found from the Missouri River to the Pacific, and he hoped to significantly extend the territory of the United States into areas under the control of European nations; he had a vision of his newly minted country stretching from sea to sea. He was also fascinated by the natural history of the lands west of the Mississippi River – there he expected to find huge ground sloths and tribes of lost 'Welshmen'! In the end none of his expectations were met but, in his choice of man to lead the exploration, he was spot on. He selected his young private secretary Meriwether Lewis to lead an expedition into the newly acquired 'Louisiana Purchase'. Lewis was a Virginia tobacco planter, like Jefferson, who had served in the military and had grown up exploring in the eastern United States; he shared Jefferson's politics and interest in the natural world. In order that he collect and observe as scientifically as possible, Jefferson sent Lewis on a crash course in scientific natural history from his colleagues at the American Philosophical Society, of which he was the president. Lewis learned how to take astronomical measurements, essential for later determining latitude and longitude, how to preserve animal skins and, importantly for us, how to press and dry botanical specimens. Lewis was given free rein to choose his team – his first choice was William Clark, an army officer under whom Lewis had served along the Mississippi River, then the western frontier. It was an inspired choice; the two men created and melded their 'Corps of Discovery' into a super-efficient unit; they were born leaders of men.

After witnessing the signing of the treaty selling the 'Louisiana Purchase' – all land to the headwaters of the Missouri River – in St. Louis, the team set off up the Missouri by boat, crossing with some difficulty lands controlled by the Lakota Sioux people – accomplished warriors who were not happy with expansion from the east. They wintered in the territory of the Mandan people, where some members of the expedition were sent back to St. Louis with collections and journals. There Lewis and Clark took

As Lewis (with the red scarf) and Clark discuss the passage over the mountains from Traveler's Rest, Sacagawea, cradling her baby born on the journey, listens and watches – her advice many times saved the expedition from disaster.

on a French trapper called Touissant Charbonneau, but most importantly also added to the expeditionary force his wife, a Shoshone woman who had been captured by Hidatsa warriors as a teenager several years before and was won by Charbonneau in a bet. This was Sacagawea, without whom the Lewis and Clark expedition would never have reached their goal. She spoke several local languages and French, and via a convoluted path served as translator to the expedition. She also was the source of much information on what to eat in this world that was unknown to the expeditionary force; while they could kill as many buffalo and elk as they liked – the plains were teeming with them – what plants were safe and nutritious was another matter. She also knew the lands of her own people, further up the Missouri River. Although the expedition team was called the 'Corps of Discovery' they were not really in unexplored territory. All the land through which they travelled had been used by indigenous peoples, many different groups of them, some at peace with each other, some engaged in constant conflict. The route they took was along well-worn trails used by these peoples to follow game and resources through the seasons. Today's highways follow the same routes.

The winter of 1804 was spent with the friendly and welcoming Mandan people in what is now the state of South Dakota, giving Lewis hope for one of his missions from Jefferson, to bring the inhabitants of the lands west of the Mississippi into the American orbit and for them to acknowledge the United States government as their new 'Great Father'. Jefferson, like so many, had horribly conflicting views on the indigenous peoples of North America. One of his ideas was that the west be given over to them, so that they could run a mercantile enterprise for the United States,

I saw *Lewisia pygmaea* at the edge of a snowbank on Beartooth Pass in southern Montana, USA – smaller than its more spectacular cousin the bitterroot, it still has the same bright pink and white flowers and fleshy leaves.

based on the fur trade. This of course did not consider either the views of the indigenous peoples, nor of the hordes of land-hungry settlers clamouring for more space, nor the unsustainability of the exploitation of animals for their fur. These conflicting pressures led ultimately to the tragic and forcible removal of indigenous peoples from their traditional lands by the end of the 19th century, and to the end of the sustainable use of the immense natural resources of western North America.

Lewis and Clark opened the way for this with their exploration, along the way relying on the groups they encountered to provide guides and information. Jefferson had particularly charged Lewis with recording what plants were used by local peoples, and this he did, sometimes with very long and detailed descriptions. Although local groups from the plains and beyond all hunted buffalo at particular times of the year, their diet was also replete with roots gathered as they travelled. Lewis's journal of 22 August 1805 from southern Montana records his first encounter with *Lewisia*:

> "… another species [*Lewisia rediviva*] was much mutilated but appeared to be fibrous; the parts were brittle, hard of the size of a small quill, cilindric and as white as snow throughout, except some small parts of the hard black rind which they had not seperated in the preperation. this the Indians with me informed were always boiled for use. I made the exprement, found that they became perfectly soft by boiling, but had a very bitter taste, which was naucious to my pallate, and I transfered them to the Indians who had eat them heartily."

The bitterroot that Lewis didn't much care for, *Lewisia*, has fleshy roots about the diameter of a pencil and a couple of centimetres long. The roots are harvested before the plant flowers, when the bitter outer skin of the root is easily peeled off. Amongst the Bitterroot Salish people of western Montana only a portion of the root is harvested, and the rest of the plant replaced in the ground so the bitterroot grounds can be visited again the next year. Women were the custodians of knowledge about root crops and harvesting, surely something that Sacagawea contributed to the expedition's success and survival. Her connection with *Lewisia* is commemorated in the species named for her in 2005 – *Lewisia sacagaweae*.

As well as recording the plants that people used, Lewis made collections using the methods he had been taught; he eventually collected the "plant of which these roots formed part" that was ultimately named *Lewisia* on the return journey the next year – the label reads "The Indians eat the root of this/Near Clark's R./Jul 1st 1806". He pressed his plants between folded sheets of blotting paper, drying them in the sun – a laborious process. That more than 200 specimens of plants collected on the expedition still survive is amazing; they represent not only species and genera that were not known to scientists at the time, but the incredible effort Lewis took in complying with Jefferson's wishes to know about the natural history of the area.

Once Lewis and Clark and their men (Sacagawea and Charbonneau had stayed with the Mandan) returned to the eastern United States in 1806, they were feted as returning heroes. Perhaps too much, at least in Lewis's case – given a prestigious post as Governor of Louisiana by Jefferson (for which he was not particularly suited as it involved politics and tact) – for he never found time to prepare his fascinating journals for publication. Meriwether Lewis tragically committed suicide in 1809. Clark fared better; married and living in St. Louis, he undertook the education of both Sacagawea's children.

Lewis and Clark had not found a water route to the Pacific, nor did the headwaters of the Missouri reach into Canada thus increasing the territory of the United States, but they had reached the Pacific Ocean and returned, by land. The specimens they collected were the true treasure though – most of what they had collected was new to science. The plants were described and given names by Frederick Pursh is his 1813 catalogue of the plants of western North America, including the genus *Lewisia*, in which he put a single species, *Lewisia rediviva*. The specimen itself has only flowers, perhaps because the root attached had not dried and was planted, hence the name *rediviva*, meaning revived! In his description Pursh records that this plant would do well in horticulture; a root was planted, but "by some accident" was killed and he never saw it flower.

After Pursh's use of the specimens in his flora, they languished in Philadelphia until the end of the 19th century, when they were 'discovered' by the botanist Thomas Meehan, who recorded many of them having been eaten by insects – the fate of herbarium specimens not well looked after. Today all these collections are digitally available – the care with which Lewis recorded the date of collection means where they were collected can be traced, thus providing a snapshot of the biodiversity of a region now completely changed by human beings.

True to Pursh's suggestion, *Lewisia* species are much valued rock garden plants with their beautiful often candy-striped flowers and succulent leaves. I wonder if that thought occurred to Meriwether Lewis as he collected along Lolo Creek at Traveler's Rest, or as he gave up his portion of bitterroot to his local hosts, who ate them "heartily"? Probably not, he was a man of the frontier and the outdoors through and through.

Linnaea

CARL LINNAEUS

Family: Caprifoliaceae, the honeysuckle family
Number of species: 1
Distribution: circumboreal

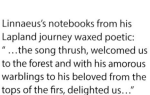

In none of the many portraits – more than 500 are known – painted or drawn of the Swedish botanist Carl Linnaeus is he without a sprig of the plant *Linnaea borealis* in his hand. It was his 'brand' and in later depictions was always accompanied by one of his books, most often the famed *Systema Naturae* – in which he classified all the plants and animals known to him. *Linnaea* appears on his heraldic coat of arms – he really identified with this little plant.

Linnaeus, born in the first decade of the 18th century, was the man who gave us the naming system we use today for all living things – what is called the binomial or binominal system, where each distinct species has a genus name and a species name – like *Linnaea* (genus) *borealis* (species). Before Linnaeus presented this way of naming in a systematic way, names were sentences, some long and some short, that described the organism in question. His two-word method was first published in the 1753 book *Species Plantarum*. Most botanical naming has been set to date from that publication, even if previous botanists had used the names before.

Linnaeus's notebooks from his Lapland journey waxed poetic: " …the song thrush, welcomed us to the forest and with his amorous warblings to his beloved from the tops of the firs, delighted us…"

Carl Linnaeus was born in rural Sweden, the eldest son of a local pastor, and is said to have had a fascination with plants from an early age. The Latinate surname 'Linnaeus' had been taken by Carl's father at university to demonstrate his learning, a departure from the then-common lack of family names – they weren't really necessary in a small, closed society where everyone knew everyone else. Carl's father signed himself Nils Ingemarsson Linnaeus – son of Ingemar, plus the Latinate Linnaeus. Carl was a mediocre student, and went to university, first in Lund and later in Uppsala. In both places he was fortunate to find mentors who helped him along the way; in Lund, his landlord Kilian Stobaeus gave him the run of his extraordinary library, and in Uppsala he met Olof Celsius, a local naturalist of repute and the professor of theology. Celsius became his first great benefactor, showing the thesis Linnaeus wrote on the sexuality of plants to the professor of medicine, Olof Rudbeck, who was deeply impressed, impressed enough to offer Linnaeus a job lecturing, even though he was just a second-year student! In Uppsala he also met Peter Artedi – a like-minded student with whom Linnaeus formed a deep and lasting friendship, sharing ideas for how to classify the world. They made a pact: if one of them should perish, the other would "consider it a sacred duty to give to the world what observations might be left behind by him that was gone." Sadly, it fell to Linnaeus to do just that when Artedi tragically perished – if they both had lived the binomial system might have been credited to them both.

Linnaeus had always gone on short field excursions to see plants in the wild, but in 1732 he set off on another sort of journey altogether. Olof Rudbeck had travelled to far northern Sweden (then called Lapland, but today by its Sámi name, Sápmi) in the early 1700s, but all his notes and

In his wedding portrait Linnaeus proudly displays his 'brand' – a sprig of *Linnaea borealis* in his hand and his arm resting on one of his many books.

material had been destroyed in the Uppsala fire of 1702; surely Rudbeck had told Linnaeus of this journey and its frustrating aftermath, sparking in the young naturalist the desire to explore for himself? So, off he went – paid for by the Royal Society of Science; not quite as much as he had hoped, but enough to get by.

Linnaeus couched his application to the Society in nationalistic and economic terms – a Swede should surely be sent so no foreigner benefited from what Sweden paid for, and surely too there were vast mineral and biological resources that could benefit the nation. And besides, he was certainly the only Swedish naturalist with the knowledge of all three kingdoms of nature. He was not really, despite later descriptions of the journey, setting off to lands completely unknown to mainstream Swedish society. Missionaries had long ago reached the farthest corners of the land, and the Sámi people had lived on the land for generations – so this was not a journey into the unknown. But for Linnaeus and for natural history, it was a voyage of discovery. In four months of travelling, he reached as far north as the Arctic Circle, collecting specimens of plants and animals, and making observations on everything under the sun as he went.

It was on this journey that he first encountered *Linnaea*. When ascending the "highest peak of Medelpad", Norbykullen, several days north of Uppsala he recorded collecting "Campanula serpillifolia" – along with some mosses and other alpine plants. In his journal of the journey, published only after his death, he never used the name *Linnaea*, but called the plant by the name it had been given by previous authors. He didn't even illustrate it, as he did with other plants he encountered on his journey

– like the bog-rosemary *Andromeda*, that he rather romantically allegorized as a maiden chained to a rock menaced by a dragon, just like the legend. His narrative throughout his diaries stresses his hardships, his discoveries and his own singular role as an extraordinary adventurer. He characterizes the region as unexplored, unknown, ready for claiming – the usual narrative of 18th century European colonial mindset. In addition to describing and recording the "three kingdoms of Nature" he also recorded in great detail elements of Sámi life. On the one hand he was in awe of their ability to live in harmony with their environment, but on the other, he later appropriated elements of Sámi culture to further his own aims. The famous portrait of Linnaeus in 'Lapp' dress is cringeworthily inappropriate – his hat is that of a Swedish tax collector – and the shaman drum he carried was almost certainly obtained illegally. This identification with the Sámi was the principal narrative that set Linnaeus up as an intrepid, adventurous natural historian, but never again was he to travel alone into the field so far from home.

Once safely back in Uppsala from his travels north, Linnaeus greatly exaggerated his journey, adding on some thousand miles to the distance travelled in a vain attempt to obtain additional funding from his sponsors. Although he had resumed teaching at the university, he was short of money, but again he fell on his feet. He was offered a job taking the son of a wealthy friend to Holland, the centre of European natural history studies at the time. With him he took the manuscripts for the works in which the name *Linnaea* appears for the first time – his *Flora Lapponica*, *Critica Botanica*, and *Genera Plantarum* – all of which were published in 1737.

It is an unwritten rule in botany that one doesn't name things after oneself – but seemingly Linnaeus did – *Linnaea borealis* is usually referred to as "the plant … that he was later to name in his own honour." Why he decided that this little plant should be his emblem is not known, perhaps its potential for use as the basis for a domestic tea industry, or its local use as a medicine that he recorded in *Flora Lapponica*, taken from Artedi's observations from central Sweden. Perhaps the idea of it as 'his' emblem came to him as he was writing up his manuscript from his great northern journey for publication. In any case, Linnaeus attributed the name not to himself, but to Jan Gronovius, who he met in Amsterdam in 1735, and who was one of Linnaeus's great admirers and supporters. In *Flora Lapponica* the name *Linnaea*, attributed to Gronovius, only appears on the illustration – in the text it is referred to as "PLANTA nostra [our plant]", along with the name that he had used in his diaries. It has been suggested that Linnaeus convinced Gronovius to receive the credit for the name, based on the assumption that the name *Linnaea* was used in Linnaeus's diaries. But it wasn't, that assumption is based on a 19th century translation that added the name *Linnaea borealis* to the text – in his travel journal Linnaeus only ever referred to the plant once, and then as "Campanula serpillifolia". The name *Linnaea* never appeared in print until 1737, in the flurry of books published once Linnaeus reached Holland, and then only in an illustration prepared after the fact. In his *Critica Botanica* of 1737 *Linnaea* is

Gift ex WTU

BM AQ: 2018-64G

DOUGLAS COUNTY, OREGON
U.S.A.

Along the Pacific coast of the United States and Canada *Linnaea* flowers are
longer and narrower than elsewhere; these populations have been recognized
as variety *longiflora*.

described with mock modesty as "a plant of Lapland, lowly, insignificant, disregarded,
flowering but for a brief time – like Linnaeus who resembles it [translated from the
original Latin]"; this is very un-Linnaean, he was usually the first to blow his own
trumpet. So maybe he did name it for himself but attribute the name to Gronovius,
or maybe he didn't and Gronovius suggested that this plant his friend was so fond of

should be called *Linnaea*. After all, Gronovius was a real admirer – in bidding farewell he wrote of Linnaeus "pre-eminent on account of the several very dangerous journeys he has undertaken for the benefit of the community – journeys during which he so felicitously investigated the three kingdoms of Nature." Whoever coined the name, *Linnaea* is forever linked to Linnaeus, both in who it honours and its publication, because of the subsequent rules of botanical naming.

Linnaea is indeed a small plant, perhaps not "lowly, insignificant and despised", but not the sort of thing that immediately catches the eye; when I have seen plants of *Linnaea,* they always surprise me, but once my eye is in, they are easy to find. The round, shiny, evergreen leaves form large mats in the dark forest understory that can be hundreds of years old. The fragrant, nodding pink flowers are paired on an elongate, thin stalk about 3 cm (2 in) long held above the leaves, hence the common name twinflower. The single species of the genus is very widespread around the northern hemisphere and highly variable in leaf shape, flower shape, and flower markings. *Linnaea* is a self-incompatible plant, meaning that self-pollination does not lead to seed set. This makes the populations very vulnerable to the effects of isolation, especially since a population of *Linnaea* may all be a single individual, given that the

Linnaea has three types of shoots; one produces clonal plants by vegetative reproduction, another creeps along the ground extending the plant horizontally and the last develops the inflorescence.

The sticky hairs on the fruits of *Linnaea* attach them to passing animals, taking the seeds away from the parent plants to establish new, genetically distinct populations.

plants can clonally reproduce. This is especially true in areas that represent isolated ranges that are thought to be remnants from the last Ice Ages – places at the southern edge of the range like the Alps, the Rocky Mountains, North Korea, and the Caucasus.

Studies in Scotland, where *Linnaea* is of conservation concern, showed that the main pollinators were flies of a variety of families; hoverflies were the most effective, but they only travelled about a metre (3 ft). Bumblebees occasionally flew longer distances between patches but were extremely rare visitors. This meant that in these *Linnaea* populations most pollen deposited on flower stigmas was from the same individual – self pollen – and was unable to fertilize ovules. So, lack of seed set was due not to lack of pollination, but to lack of suitable mate availability within patches – the patches, though large, represented very few individuals. Although flies were efficient at visiting flowers and transferring pollen, they didn't fly far, thus creating a barrier for pollen exchange between different genotypes of the plant. If no seed is set to be taken to new areas for colonization – the fruits encased in a covering of sticky hairs and attach to animal fur easily – then populations of *Linnaea* only spread clonally, making them vulnerable to random events and environmental change, thus to local extinction. So, despite its huge range all around the northern hemisphere and its conservation status of Least Concern based on this range, the population dynamics of *Linnaea* mean it should certainly merit our care and attention. Linnaeus certainly cared for this little plant, even if it was in a self-serving way.

Magnolia

PIERRE MAGNOL

Family: Magnoliaceae, the magnolia family
Number of species: *c.* 225
Distribution: worldwide except Africa

Magnolia flowers were previously thought to be 'primitive'
flowers, like those of the earliest angiosperms, or flowering plants.
Their many separate parts of indeterminate number arranged
in spirals suggested to early botanists an early state in evolution that
preceded the evolution of flowers with fixed numbers of sepals and
petals, stamens differentiated in form, and carpels fused into a single
structure, the ovary. Phylogenetic research coupled with the discovery
of extraordinary flower fossils, however, tells a different story. Rather
than the large fleshy blousy magnolias being the earliest flowers,
botanists now understand the earliest flowers to have been very small.
Reconstructions of a hypothetical ancestral angiosperm flower
indicate that it likely had the flower parts in whorls or rings,
with undifferentiated tepals, stamens and separate carpels.
Magnolias represent a derived lineage with more flower parts
than those early angiosperm flowers that are arranged spirally,
rather than in distinct whorls, a physical result of packing of

Magnolia fruits are cone-like structures made up of separate carpels, each a follicle with a single seed often encased in a red aril.

more flower parts optimally. The large open magnolia flowers with their fleshy, sturdy
parts are magnets for beetles, who come in search of pollen – a rich source of protein.
In their bumblings around in the flowers, the insects become covered in pollen and
come into contact with the many stigmas of the ovule-bearing carpels, thus effecting
pollination. Magnolias are beetle-pollinated, their flower parts are tough in part to
withstand the actions of these insects.

The first magnolias seen by European botanists were from the Caribbean – the
name *Magnolia* was coined by Charles Plumier for plants from Martinique. The name
honoured his contemporary the French botanist Pierre Magnol of Montpellier, "Among
the great and meritorious botanists, who in early years learned medicine and enlarged
this to illustrate botany, whose works merits recognition for greatness." Plumier
cited one of Magnol's last works, the *Hortus Monspeliensis* of 1697, as the reason for
his greatness, but more than this catalogue of the plants growing in the Montpellier
botanical garden was the reason for Magnol's high standing amongst the botanists of
his time and later.

Pierre Magnol was a member of a family of apothecaries in the town of Montpellier, in what is today the Languedoc region of southern France. Montpellier was an intellectual and cosmopolitan city where the influence of Arab medicine and natural history was celebrated and valued. It was a centre for the spice trade, and the mixing of cultures and ideas formed part of the fabric of society. The school of medicine attracted students from all over Europe and beyond. A botanic garden was founded in the city in 1289, devoted to the teaching of medicine and pharmacy – botany and medicine were intricately linked, to be a doctor you must know your plants! Magnol attended university in Montpellier and qualified as a doctor in the mid-17th century. He also travelled widely in the region and as far as the Pyrenees and Alps to study plants in their native habitats. He was widely considered a brilliant student and, with his qualifications and knowledge, was an ideal candidate for a position as professor of medicine at the university. But the France of the time, like much of Europe, was fraught with religious conflict and discrimination. The Reformation had arrived in Montpellier in the early 17th century, the city's population was largely Huguenots – Calvinist Protestants – rather than belonging to the Catholic state religion of the time. Magnol was a Protestant, and the professorship went to the Catholic candidate – despite the laws granting Protestants full civil rights including that of holding public office.

This lack of promotion did not deter Magnol, however, he continued to teach medical students in the science of botany, leading them on excursions in the local area to learn plants in the field and as living organisms. From the observations and collecting done on these excursions he developed the *Botanicum Monspeliense*, in which he listed the plants of the region around Montpellier in alphabetical order with short descriptions of their form and colours, plus notes on how they were used, where they grew and when they bloomed. This book doesn't really look like the floras or plant guides we have today. For a start, Magnol was writing in the late 1600s, before the advent of binomial names for plants we are familiar with today. He used Latin polynomials, essentially short sentences of several descriptive words – the woody nightshade known today by the

The artist Georg Dionysius Ehret walked the distance from Chelsea to Fulham Palace every day to see *Magnolia grandiflora* bloom for the first time in London.

scientific name *Solanum dulcamara*, is listed as "SOLANUM scandens seu Dulcamarae" – and the plants are listed in alphabetical order, not by similarity or relatedness.

Today's floras are usually organized by plant family – so all the magnolias, for example would be found together, facilitating identification and differentiation. This idea of the plant family comes from Magnol, although he didn't use it in his *Botanicum Monspeliense*. About the same time as Magnol was updating the flora in the late 1680s he also published one of his most influential works, the *Prodromus historiae generalis plantarum* [a general history of plants] in which he divided all the plants known to him into 75 tables or groups, based on associated common features, like form, number of parts etc. For these groups he used the term "familias" or families "by comparison with the families of mankind."

Magnol also put his finger right on the nub of why a reliance on single characters are not necessarily helpful for grouping plants – they sometimes conflict with each other. Until the concept of common descent was well-established, efforts to come up with a natural classification all ran into this issue – that affinities "strike the sense but cannot be expressed in words." Today's cladistic methodologies, where emphasis is placed on those characters that are shared from common ancestors, allow us to look at these conflicts and begin to explain them biologically.

With the publication of the second edition of the flora and of the *Prodromus*, Magnol finally achieved the standing he deserved – but not without cost. In 1685 the Edict of Nantes – granting status to Protestants – was revoked by Louis XIV, making persecution, emigration, or conversion to Catholicism the only routes open for Protestants to remain in France. Many Huguenots fled France seeking refuge in neighbouring Protestant countries like Switzerland, the Netherlands and Britain. Magnol chose conversion, and became a Catholic, opening the way for his promotion to a public post at the university. Ten years later he received a royal commission as professor of botany at the Royal Botanic Garden in Montpellier – job security at last.

Despite his admiration for Pierre Magnol and his work, Linnaeus did not use Magnol's system of families in his great *Species Plantarum* of 1753, the work from which the binomial system of plant naming dates. He did, however,

Magnolia campbellii, whose dinner-plate size flowers emerge long before the leaves, cloaks Himalayan hillsides with pink in the early spring.

cite *Magnolia* as an example of how generic names based upon people should reflect the characters of both plant and person saying: "*Magnolia* is a tree with the most splendid leaves and flowers named for the most splendid botanist." Adopting Plumier's name *Magnolia*, Linnaeus grouped all of the different types known to him into a single species – *Magnolia virginiana*. This included the species seen by Plumier – now known as *Magnolia grandiflora* – that stately tree lining the avenues leading to the plantations of the southern United States of America, sites of so much luxury and hardship side by side.

Tightly clustered carpels and stamens sit at the centre of the *Magnolia virginiana* flower; common names for this species are sweetbay or laurel magnolia, from the spicy scent of the bark.

Today we know about 300 species of *Magnolia*, all grouped in a single genus following intensive study using genomic tools. For a long time, classifications of the genus varied between this grouping and one where up to 17 different genera were recognized – in part due to divergence of opinions between taxonomists in different countries. Now opinion has converged on a large, inclusive *Magnolia*, with fifteen subdivisions recognized, many of which correspond to the previously recognized genera. *Magnolia* is one of the genera that exhibit one of the most fascinating biogeographical patterns in flowering plants, one that intrigued Charles Darwin and the Harvard botanist Asa Gray in the mid-19th century – the disjunction between eastern Asia and eastern North America.

The origin of this distribution pattern, with species in the same genus found in centres of diversity in the widely separated areas of Asia and North America was put down to the existence of a more widespread and disrupted north temperate flora in the past. The advent of knowledge about how the continents have moved and changed positions through geological time has completely changed how we think about plant distribution patterns on the Earth's surface today. The eastern Asian–North American disjunction has been explained by a variety of factors, widespread floras across the supercontinent of Laurasia with extinction in the intervening areas, plant movement across both Atlantic and Pacific land bridges during the maximum development of the thermophilic (warm-loving) deciduous forests in the Tertiary, and long-distance dispersal over oceans. The pattern we currently observe, and that Asa Gray remarked on, is the result not of one process, but the conjunction of several across time periods ranging from the pre-Tertiary to the Quaternary.

collections consisting of dried plants mounted on paper; about a quarter of new species were described based on specimens collected more than 50 years before. So it's pretty exciting when a new genus is found in the field!

The plants had a strange combination of characteristics, they fit into none of the recognized genera of the family, but shared some characters with the genus *Lopezia*, another Mexican endemic. *Megacorax* didn't quite fit there either, so its evolutionary relationships were still unclear. Subsequent work with DNA sequence data showed that *Megacorax* is sister to *Lopezia*, and the two genera form a clear monophyletic group that is itself sister to most of evening primrose diversity. Questions still remain though, are there other species of *Megacorax*? How common or rare is it? Describing a new genus is only the very first step to understanding it.

Although only two populations of *Megacorax* have been found, it is likely to be more abundant in the area – there are no roads in the poorly known mountains of the Sierra de Coneto in the floristically diverse state of Durango. Durango has the richest flora of

The flowers of *Megacorax* are typical of the primrose family, with four petals and four sepals, but its possession of eight stamens in two whorls is one of the features that make it unusual in this family.

all the states of northern Mexico, in part due to its central position, with the arid the Chihuahuan Desert bounded on either side by the main mountain ranges of the Sierra Madre Occidental to the west and Sierra Madre Oriental to the east. whose habitats range from pine and oak forests to tropical rain forest to arid deserts. The Sierra de Coneto is an isolated volcanic range on the eastern flank of the Sierra Madre Occidental, with pine and oak forests interspersed with open rocky slopes with very acidic soils. It was under shrubs in these rocky areas that *Megacorax* was found during vegetation surveys. Once its novelty was confirmed, botanists returned to the locality and found that *Megacorax* was relatively common, it was even a dominant part of the low shrubby layer. Its perceived rarity may just be due to lack of collecting in this part of Mexico.

As with *Megacorax*, our knowledge of plant distribution and abundance is far from complete. Plant collecting can take many forms, from an exploration of an area previously not well-sampled with collections of all plants being made, to a targeted trip to find one particular plant that is needed for a DNA sequence analysis to everything in between. Collecting and observing plants in the field, seeing them as organisms in their environment, not just DNA sequences or sets of characters, is something that has been central to Peter Raven's botany from the beginning.

Peter Raven was born in the bustling Chinese city of Shanghai in 1936. His father worked in the family bank, so the family was part of the flourishing ex-pat financial community. But disaster struck, the bank began to fail as the aftermath of the Depression hit, and his uncle, who was the manager, was accused of financial mismanagement and jailed, first in China then in the United States. So, Peter, then about a year old, and his parents moved back to California, where both his parents' families were located; this move came just in time before Japanese armies invaded and occupied Shanghai, along with much of the rest of China.

Growing up in San Francisco, Peter Raven became a member of local natural history societies, meeting and interacting with many of the famous California botanists of the time, like Alice Eastwood and John Thomas Howell of the California Academy of Sciences. His passion for plants and keen interest in the field meant he very soon became a mainstay of the California Native Plant Society and began making herbarium collections of his own. Like many a young naturalist, his interests wavered between beetles and plants, but he came down on the side of plants, much to botany's benefit. Describing his impulse to collect in those times, Peter said, "I was collecting plants because I wanted to know them, all of them."

Passion for the California flora followed Peter through his university years, as both an undergraduate and then as a PhD student entering his doctoral studies "more as a lover [of plants] than a scientist." His work for his doctorate was on the evening primrose family and, true to form, Raven studied these plants intensively in the field. He also began a paradigm-changing interest in plant-animal interactions, that was later to develop into a collaborative body of work with Paul Erhlich on co-evolution – how plants and animals interact and drive change in each other.

Peter Raven (b.1936) in a colony of the bamboo *Yushania* in the mountains of Taiwan, China. His love of plants has meant he always wants to experience them first hand, even on editorial visits.

Broadening his world experience after gaining his doctorate, Raven came on a post-doctoral fellowship to London, where a fellow young botanist described him as the "California whizz-kid", memorable for "the loud clackety-clack of his manual typewriter echoing through the normally deathly silence of the main herbarium, not to all the inhabitants' pleasure." Raven's work ethic was, and still is, legendary. Using the time in Europe to consult old collections not available in the United States, and collecting and examining European evening primroses in the field, Raven consolidated himself as the up-and-coming expert in this family.

Soon after his return to California, he was offered a position at Stanford in the emerging field of evolutionary biology – at the tender age of 26 years old – he was not much older than many of his students, so as he said "I learned to present myself as a personality, to entertain as well as educate … It was important to be present and aware." His lectures were in the classroom and, of course, in the field. At Stanford, his partnership with Erhlich blossomed, born from a mutual fascination in how things worked in nature. Here too he took his first trip to the tropics and began a life-long commitment to concentrate on conservation and sustainability, "I knew I had to contribute whatever I could do the effort to stave off the massive losses of biodiversity we were beginning to see around the world." This was the late 1960s.

Just before going to Stanford, he had another epiphany that would change and broaden his focus from that of an academic botanist delving ever deeper into his or her specialist subject. He realized that he could make an impact in and on science by helping others – he began too a lifetime of encouraging and helping others in the grand task of cataloguing life on Earth. A move to the directorship of the Missouri Botanical Garden in St. Louis in 1971 made all of those dreams come together. At his new base he took every opportunity to strengthen and increase the Garden's effort in plant inventory, especially in the more poorly collected and understood parts of the world. He may never have encountered the suggestions of Alfred Russel Wallace from his 1911 book *The World of Life*:

> "There must be hundreds of young botanists... who would be glad to go collect, … if their expenses were paid... Here is a great opportunity for some of our millionaires to carry out this important scientific exploration before these glorious forests are recklessly diminished or destroyed – a work which would be sure to lead to the discovery of great numbers of plants of utility or beauty, ..."

But Raven's thoughts ran along the same lines, and there were many young botanists eager to go and collect in the "these glorious forests", meaning the tropics – I was one of them and the experience profoundly influenced the rest of my career. Raven's way was unique though, in that his method involved strengthening local institutions and botanists, and even institution-building in some of the most biodiverse places on the globe, like Madagascar and Peru. His conviction that together we can do so much more than if we work alone influenced so many, those who were directly employed by the Garden as collectors, his own students and the many others whose work he helped or whose lives he touched.

That focus on cataloguing plant diversity was not a simple stamp-collecting enterprise, Raven's impetus throughout was on the conservation of what were increasingly rapidly disappearing natural habitats. His advocacy for the planet brought the issues of tropical deforestation and environmental change through human action to a wider public. He tirelessly convinced those "millionaires" to support study of plants and their conservation before the forests disappeared forever.

Plants have been at the centre of Raven's life – how to know and understand them, how to preserve them, how to make them central to all our lives. His lifelong study of the Onagraceae – evening primroses – is fittingly commemorated in the still poorly known genus *Megacorax* – whose name, in his words "means, embarrassingly, 'great raven'!" - about which there is still so much to discover.

Peter Raven – that *Megacorax* – has led by example, by helping and supporting others, by showing that science and sustainability cannot be viewed separately, and by continuing to speak up for the continued scientific exploration of the world around us. His ability to connect – topics, people, initiatives – is second to none. I leave the last word to him: "We cannot afford to use our record of progress as an excuse for complacency and inaction. We can't just assume that 'they' will work out solutions to all these [societal and environmental] problems. We – all of us collectively – can change the world for the better and we must begin now."

Meriania

MARIA SIBYLLA MERIAN

Family: Melastomataceae, the meadow beauty family
Number of species: *c.* 116
Distribution: Mexico, the Caribbean islands and South America

Since the time of Linnaeus botanists have primarily relied on the shape and colour of flowers and fruits – the reproductive parts of the plant – for identification. Identifying sterile plants from just their leaves can be extremely challenging, especially in the tropics. Ecologists working in forest plots in the tropics seeking identification of branches without flowers or fruits can be met with sighs of exasperation from more conventional botanists; DNA sequencing can help to a great extent, but the ability to rapidly identify plants even down to family in the diverse tropical rainforest is a true skill. One plant family though is an easy one to identify from leaves alone – the Melastomataceae. Leaves of melastomes (as botanists shorten the name) are easy to spot – they almost always have a few main veins from the base, with ladder-like secondary venation connecting these. This regular pattern means this is one of the first flowering plant families botanists learn in the tropics!

Melastomes are mostly a tropical family, a few species extend to the temperate zone, like the meadow beauty of the southern United States of America – *Rhexia virginica*. Two-thirds of the genera of melastomes are from the American tropics, including *Miconia*, now one of the largest genera of flowering plants with over 2,000 species. The flowers of melastomes are as distinctive as the leaves; the anthers are often horned or have strange-looking appendages and open by tiny pores at the tips, rather than opening as do the anthers of many other flowering plants via a slit along the side. Poricidal anthers of melastomes are indicative of a specialized bee pollination system – called buzz pollination (see *Sirdavidia*), so named because audible buzzing noises can be heard during insect visits. Most melastomes have buzz-pollinated flowers, but *Meriania* has diversified to attract vertebrate pollinators. Many species, to be sure, still are buzz-pollinated by bees (like *Meriania speciosa*), but the stamens in others are highly modified and arranged in ways that facilitate pollination by vertebrates. Stamens in some produce nectar rewards, and others have bulb-like appendages that are food bodies eaten by visiting birds. *Meriania* species that produce nectar as a reward are often visited by birds like hummingbirds in the

The opposite leaves with their distinctive ladder-like venation make the large-flowered Andean tree *Meriania speciosa* instantly recognisable as a member of the Melastomataceae.

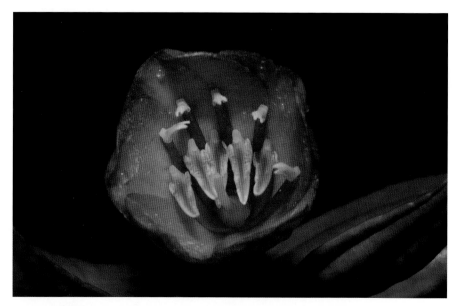

Meriania angustifolia from Cuba is not buzz-pollinated, but instead has a 'mixed-vertebrate' pollination system with a nectar reward; pollen shakes out of the pores like salt when an animal touches them reaching for the sweet stuff.

day and by bats or rodents at night – a remarkable way in which evolution may have driven exploitation of flower visitors 24/7.

Meriania is a typical melastome tree or shrub with ladder-like leaf venation and variously horned anthers of different shapes in the same flower. Unlike many other melastomes though the flowers are large and showy – in most of the species the anthers, with their projections or appendages, are held to one side of the flower, making it bilaterally symmetrical. The genus *Meriania* was named from plants collected in the Caribbean by the Swedish botanist Olaf Swartz in honour of the entomologist and naturalist Maria Sibylla Merian, whose paintings and collections of Surinamese plants and insects were the first to depict tropical plant-insect interactions. Swartz's dedication recognized Merian's botanical as well as entomological contributions:

> "In memory of Maria Sibylla Merian, born in Frankfurt am Main, who journeyed to Suriname, and whose splendid and beautiful book entitled *Metamorphosis of Surinamese Insects* described and depicted plants which are only now becoming known to botanists."[Translated from the original Latin.]

Swartz didn't name *Meriania* for Maria Sibylla Merian because she had collected or depicted it, but for her profound impact on our knowledge of tropical plants. Her work was ground-breaking in many ways, not just botanically – she was the first female European naturalist to independently study nature *in situ* in the tropics, all this in the 17th century. She was born in 1647 and grew up in a family of engravers and artists;

her father was an engraver, and her stepfather a noted flower painter. So publishing and artistry were part of her everyday life from an early age. She assisted her stepfather, Jacob Marrel, in painting flowers for the lucrative Dutch market and as a young girl became interested in insects. Starting with silkworms, she realized that they turned to moths, so she began to describe and depict their life cycles: "I realized that far lovelier butterflies and moths emerged from caterpillars other than silkworms. This inspired me to gather all the caterpillars I could find in order to observe their transformation. That is why I withdrew from society and engaged in this study. In order to practice the art of painting and to be able to draw and paint them from life, I painted all the insects I could find ..."

In the mid-17th century, the phenomenon of insect metamorphosis was still not well-understood. Today we know that various orders of insects – flies, bees, wasps, butterflies and moths – undergo dramatic transformations through their life cycles, transforming completely in shape from larvae to adult. The Greek philosopher and naturalist Aristotle characterized all animals as viviparous (giving birth the live young), oviparous (whose young hatch from eggs), or vermiparous (that arose as worms spontaneously from rotting matter). Maggots and caterpillars – the larvae of flies and moths – were vermiparous, the product of spontaneous generation. Two years after Merian was born, the British scientist William Harvey, best known for his description of the human circulatory system, disputed the Aristotelian idea of spontaneous generation and argued that maggots and caterpillars were actually eggs – making these insects oviparous. He called the mobile, feeding larvae 'imperfect eggs', suggesting that the static, immovable pupal stage was the 'perfect egg' stage of development. The Dutch scientist Jan Swammerdam, whose work Maria was familiar with, disputed this idea of insect development, suggesting instead that the transition from larva to adult was a distinct series of changes, and was a process of growth. So, at the time Maria began to investigate caterpillars of silk moths and other moths and butterflies – this transition from egg to caterpillar to pupa to adult flying insect was still under considerable debate.

Merian married her stepfather's apprentice at the age of 18 and soon had two daughters. The family moved to Nuremberg, where she gave drawing lessons to the daughters of rich merchants and noblemen, allowing her access to gardens where she continued to investigate the life cycles of insects. She also earned money by producing popular flower paintings, engravings and embroidery patterns of plants and insects. She was

Maria Sibylla Merian (1647–1717) was recognized as a scientist in her lifetime, unusual for a woman. Her standing as an expert is emphasized by the natural history objects surrounding her in this portrait.

Merian's depiction of plants with their specialist insects was innovative. Here hawkmoths feed on the leaves of a pepper plant that is somewhat fanciful, with several types of peppers from the same plant – not as accurate as her insects!

unusual in that she not only depicted insects in association with flowers, but studied them in detail, collecting caterpillars and rearing them on to adulthood as butterflies or moths. This might seem easy, but to rear a caterpillar to adulthood not only requires patience but also knowledge of what leaves it eats, how often to change the food plant and where to find copious supplies of food for these voracious creatures. This is dedicated work.

Merian published the first volume of her book on insect life cycles, *Der Raupen* [The Caterpillars], in 1679, just after the family had returned to Frankfurt. Her marriage was not a happy one, and when her stepfather died, she left her husband and went with her daughters to live with her widowed mother. All four women soon joined her younger stepbrother Caspar in a community of Labadists at Waltha Castle in the Dutch province of Friesland. The Labadists were a Protestant religious community movement founded by the French cleric Jean de Labadie, among whose tenets were communal ownership, the absolute equality of men and women, belief that all people were priests, and that marriage was dissolvable if one partner did not adhere to community values.

After her mother's death, Maria and her daughters moved back to Amsterdam, where she earned her living trading in natural history specimens and with flower paintings. Her interest in going to the tropics was piqued by the insects she saw in the collections of wealthy burghers of Amsterdam, a "long dreamed of journey to Suriname". Why Suriname? In the 17th century this part of northern South America was under Dutch control, mainly for sugar production. In addition, the governor of Suriname was one Cornelis Van Aerssen van Sommelsdijck whose family had given Waltha Castle to the Labadists. The stars all aligned and, by selling paintings and specimens, Merian financed passage for her and her unmarried daughter Dorothea to the tropics. That she, an unmarried 52-year-old woman accompanied by her daughter, was given permission by the city of Amsterdam to travel to Suriname with the express purpose of studying natural history says a great deal about the regard in which Maria Sibylla Merian was held at high levels of society. They visited plantations operated by Dutch settlers worked by legions of enslaved human beings; Maria wrote of the brutal lives of the enslaved with distaste. She and Dorothea joined the elite Dutch Labadist community at the sugar plantation of La Providentia, the southernmost Dutch outpost in Suriname, where the two women collected caterpillars and insects, as well as observing frogs, toads and snakes. Much of their work was focused on the life cycles of these animals – they were collecting caterpillars and rearing them to adulthood – connecting the larval and adult stages of insects. This was what Maria had done earlier in Europe, but in the tropics the task is even harder. Food plants can be hard to find, the heat means leaves wilt as soon as they are picked, wasps and ants eat the developing caterpillars, and parasites lay eggs on the caterpillars so that what comes out of a pupa is not a butterfly, but a parasitic fly. With a colleague I once reared caterpillars in the tropics – the rounds of feeding, cleaning, and gathering food were unending. We kept our developing caterpillars in little plastic pots, Maria and Dorothea kept theirs in wooden boxes; it must have been much harder to keep them from drying out or getting mouldy.

Their stay in Suriname, originally projected to be for five years, ended after a little more than a year with Maria's illness, probably from malaria. Even though she was ill, Merian continued to work on the voyage back to Europe – moths emerged from pupae on the way back! They returned to Amsterdam in late 1701, with collections of animals, both living and preserved, that were offered for sale to collectors of natural history specimens. She corresponded with many of the scholars of the time, who were anxious to hear of her discoveries. With Dorothea, she prepared her work for publication – they both painted and did some of the engravings, and with funds from those who subscribed in advance to the finished work, employed engravers to assist them. The work was advertised in England as by "That curious person *Maria Sibilla Merian*" – I suspect that curious in both senses was meant, this was a book by a woman, unusual in the early 18th century, who was definitely curious about the natural world.

The result was the "splendid and beautiful" book cited by Swartz, *Metamorphosis Insectorum Surinamensium*, published in both Dutch and Latin in 1705. The illustrations are masterpieces of natural history observation and are the first to depict ecological interactions in accurate detail – none of these were done in the field, they were all done from notes subsequently, so some of the details are a bit askew. Some of the caterpillars are not on their host plants, some of the insects are a bit fanciful – but nevertheless, this was tropical ecology depicted by one who had observed it carefully in the field, not from mere dried or pinned specimens. Her observations and illustrations were discussed and used throughout the scientific world of the day – decades after her death in 1717 the Swedish botanist Linnaeus used her illustrations as evidence of new plant species from tropical America, and they were also the inspiration for other explorers and documenters of nature who came after her. Her depiction of plants with the insects associated with them brought not just new plants to light, as Swartz said in his dedication of *Meriania*, but the centrality of interactions in nature to the study of life on Earth.

Quassia

KWASI

Family: Simaroubaceae, the tree-of-heaven family
Number of species: 2
Distribution: tropical America and Africa

The Swedish botanist Carl Linnaeus, who gave us the binomial system of naming, also laid out a series of 'aphorisms' or rules for how to do botany in what he considered the correct way. One of his many rules for naming plants was that generic names commemorating botanists should be carefully given. So it is unusual that the generic name *Quassia* honours not a famous European botanist but a man of African descent who had previously been enslaved. In the original description of *Quassia* no mention is made of 'Quassi' for whom the genus is named, but in a dissertation of one of Linneaus's students, Carl Magnus Blom, written the year after, the plant and its uses are treated in detail, with mention of the "servant Quassi" as the source of botanical specimens.

Quassi, better known as Kwasi, was enslaved in West Africa and, as a young man, brought to Suriname in the early 1700s to work on the plantation called Niewe Timotibo – hence another of his names, Quassie van Niewe Timotibo. Whether he received his name in Africa, where Kwasi indicates a child born on Sunday in the Akan language of Ghana, or in Suriname is not known. He worked as a scout for plantation owners, seeking out 'Maroons' – enslaved people who had escaped the plantations and set up independent villages in collaboration with indigenous peoples in the interior of Surinam and who jealously guarded their freedom and independence. He became known as a helper of the colonizers, and had bestowed upon him another name, Kwasimakumba – Kwasi the white man – by members of Maroon communities, whose oral traditions still portray him as a spy and a traitor. His forays into the interior brought him into contact with the indigenous peoples of this part of northern South America, from whom he learned much about the medicinal and magical uses of plants. He was a larger-than-life character, respected and feared in equal measure. A description of him written shortly after his death gives a flavour of this:

> "The Negro Quassy, who gave his name to the wood which he discovered and who made himself famous in Suriname through his reputed sorcery, engaged for a long period of years the minds of most of the colonists. He was often employed to go to the plantations to discover the poisoners among the Negroes; he was consulted on all sorts of illnesses .... The

Quassia amara is instantly recognizable not only by its bright red flowers, but also by its striking leaves with a winged central axis and red-tinged veins.

S.Edwards del   Pub. by W.Curtis, S.t Geo: Crescent Nov. 1.1800   F.Sansom sculp

such compound. These substances are also often used as insecticides and vermifuges – hence their efficacy in stomach complaints that were often caused by parasites of various sorts. The value of such plants, both to those who traditionally used them and to European colonizers, was huge. *Cinchona* bark, for example, played a major role in power plays between European nations in the 17th and 18th centuries – its supply was tightly controlled by the Spanish, and alternatives were sought everywhere. One of these was quassia wood.

One of the many chemicals in the wood that gives quassia its bitterness is called quassin, one of the bitterest substances known in nature. Extracts of the compounds are made with water, and wood can apparently be used many times over. Traditional usage in the Guianas is as a tea made from leaves, but the export product so prized in European markets was the wood, much more easily transported. In the dissertation on 'Lignum Quassiae' Linnaeus compared the efficacy of quassia wood to 'Cortice Peruviana' (cinchona bark) and also to gentian – a European plant well-known for its bitter qualities; suggesting that based on his experiments with patients with various ailments that the drug was well worth acquiring and was one of the great new exports from the Americas.

But quassia wood never had the market share that did Peruvian bark (cinchona or quina) as a cure for malaria. Celestino Mutis, a Spanish botanist based in 'Nueva Granada' (today's Colombia) suggested that a bark from the Guianas, most probably *Quassia amara*, was far inferior to the true quina, thus not worthy of consideration. He said "... la Corteza de Guayana no es Quina" [the bark from the Guianas is not true quina]. Science and empire were inextricably intertwined as European powers sought to gain control over large swathes of the rest of the world. Once both the British and the Dutch wrested control of the cinchona trade from the Spanish in 19th century that became the antimalarial drug of choice for European colonists; quassia remained important only locally.

Quassia appeared in the pharmacopoeias of the United States and various European countries in the 19th century, always as a herbal medicinal tonic; more recently the plants have been used in medical trials for many diseases, including several cancers. Chemicals extracted from *Quassia* have even been used as an insecticide in organic agriculture, with somewhat unbelievably no ill effects on non-harmful insects. As organisms harmful to human activity, either through causing disease or harming our food sources, develop resistance to the chemicals we use to control them, the natural world has become a fertile natural laboratory for finding new methods of control.

In the early part of the 21st century studies of traditional treatment of malaria in the Guianas revealed that *Quassia amara* was the most commonly used plant for both cure and prevention, and a patent was taken out on one of the compounds extracted from it - simalikalactone E – one of the many bitter quassinoids so long used traditionally. The filing of the patent sparked a controversy in French Guiana which became known as 'l'affaire kwasi' – after the local common name for quassia – revolving around the use of traditional knowledge for monetary gain and violation of the principles of the Nagoya Protocol, one of the global agreements associated with the Convention on Biological Diversity that requires benefit sharing with the originators of knowledge later used commercially. Legal battles concerning this patent continue to this day.

Prior and informed consent for use of knowledge given to researchers is today something we take for granted and try to adhere to, but this has not always been the case. Too often through scientific history the information shared by others is left in the shade as single individuals take credit for discovery or for significant advances. As in the case of *Quassia* and the 18th century discovery of its use for treatment of malaria, the complex interconnectedness of human interactions can leave tangled threads that are next-to-impossible to unknot. Science and discovery are collective enterprises, we all depend upon the insights of others.

The red tubular flowers of *Quassia amara* hold copious sweet nectar at the base of the tube, hidden from all but the hummingbirds whose long beaks and tongues can reach it.

Rafflesia

THOMAS STAMFORD RAFFLES

Family: Rafflesiaceae, the rafflesia family
Number of species: *c.* 30
Distribution: Southeast Asia (incl. Philippines)

Imagine walking through dense tropical rainforest and coming across something that defies the imagination – it happens a lot in the rainforest – but Joseph Arnold, the young medical man who had accompanied the recently knighted Sir Thomas Stamford Raffles and his new wife to Bencoolen [today's Bengkulu], Sumatra, saw something that stopped him in his tracks. He wrote to a friend:

> "I rejoice to tell you that I happened to meet with what I consider as the greatest prodigy of the vegetable world. [...] ....To tell you the truth, had I been alone, and there been no witnesses, I should I think have been fearful of mentioning the dimensions of this flower, so much does it exceed every flower I have ever seen or heard of..."

The flower the young Indonesian man with them found was "a full yard" (more than a metre) across, weighed approximately 7 kg (15 lbs), and came straight out of the ground – it "had precisely the smell of tainted beef" and was buzzing with flies. The structure was thick and succulent, and the flower parts were confusing indeed. Arnold didn't see any leaves, so assumed they must come out at a different time of the year. They preserved the flower and two large buds in spirit to examine later.

Sadly, Arnold died shortly after this trip to the area around Pulo Lebar village in southwestern Sumatra. In August of 1818, Raffles sent Arnold's letter, with the preserved flower and buds, to Sir Joseph Banks. Robert Brown, who was Banks's librarian and who went on to be the first Keeper of Botany at the Natural History Museum, London read the paper describing the "Great Flower of Sumatra" shortly

The fruit of *Rafflesia arnoldii* is large and filled with thousands of tiny seeds, insurance by numbers since a seed must connect with a host to germinate and the chances of that are pretty slim.

after Banks's death in the summer of 1820. Brown, who had an interest in parasitic plants, realized that the flower – named *Rafflesia arnoldii* by him in honour of two of those who were present at its finding – was that of a parasite; there were no leaves, nor would there be any. The buds and flowers emerged directly from the roots of the host plant; he could tell by looking carefully at cross-sections of the area where the bud emerged from the woody root. He also attempted to make sense of the very confusing structure of the flower; realising it was unisexual "which I did not suspect" and only had stamens and anthers (male parts), he expressed a wish for a female flower. He got that wish a decade later, as new specimens came in from Raffles and others published on plants found in other forests of Southeast Asia.

Rafflesia has without question the largest flowers of any plant, but there is considerable variation in size across the genus. Brown puzzled over its relationships to other flowering plants – the structures were just very odd. But as new species were collected and he examined new material, he concluded that they were a "natural family" – he called it the Rafflesiaceae. But what were these strange plants related to? Fast forward to the 21st century and the advent of DNA sequencing which allows

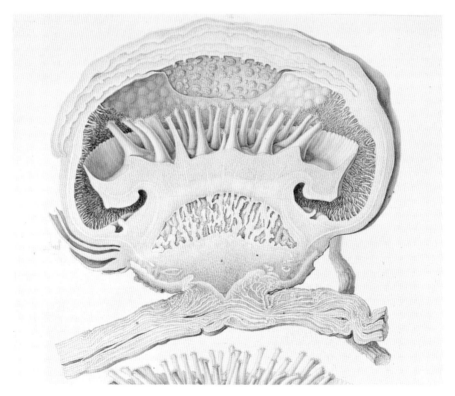

Robert Brown's careful dissection of the material sent to him by Raffles clearly showed the parasitic nature of the plant attached to the host plant root with special organs called haustoria.

The flies clustering around this opening flower of *Rafflesia arnoldii* have been tricked by its smell into thinking this is a good place to lay their eggs – good for the flower, bad for the flies.

scientists to use a new set of characters to look at evolutionary relationships. These data showed *Rafflesia* and its relatives to be closely related to the spurges – the family Euphorbiaceae – with very small flowers! Examination of the rate of floral size evolution showed that the rate of flower size evolution was extremely high in the stem lineage (the branch leading to all extant members of the family) of the Rafflesiaceae, but once the component species and genera of the family began to differentiate the rate is the same as that for the Euphorbiaceae.

Those flies buzzing around the flower that Joseph Arnold saw were serving an important function for these huge blooms. *Rafflesia* is fly-pollinated – the stinky odour of rotten beef attracts carrion flies who are looking for protein or a place to lay their eggs. The brownish purple colours of *Rafflesia* flowers, speckled with white mottling, also look at bit meat-like and their huge size is probably a super-attractor! On male flowers flies crawl into grooves on the undersides of the spiky disc in the middle of the flower. It's a tight fit, but once they get in far enough, they must back out and by then they will have brushed against the anther and have a blob of pollen on their thorax. Then – if the plants are lucky – the fly goes to a female flower and, in their searching around, brush against the stigma, which is in a similar position in the flower. That this even happens is extraordinary – considering that a fly needs to go from a male flower to a female one, in that order and often over long distances, to effect pollination. Because the sexes are separate in *Rafflesia* there is no opportunity for self-pollination. No wonder these plants were thought rare even in Raffles's time, not only is pollination difficult but they are also dependent upon another plant, their host.

Parasitism is an unusual lifestyle for plants but has evolved several times independently in the flowering plants. Parasitic plants lack chlorophyll and the machinery for photosynthesis; they rely on their host plants for their nutritional needs. These parasites attach to the vascular tissues of a host plant and tap into the pipes running water up to leaves and sugars produced by photosynthesis running back down. *Rafflesia* is what is termed a holoparasite – it has completely lost the ability to make its own food; some other parasitic plants are not so dependent and have retained a bit of ability to photosynthesize, but still obtain most of their nutrition from their hosts. *Rafflesia arnoldii* is parasitic on a huge woody vine in the grape family, *Tetrastigma rafflesiae* – the host plant named after its parasite! Other species have different hosts, but all in the grape family, and most of them *Tetrastigma*. These are obligate hosts; *Rafflesia* is found nowhere else but attached to these woody vines in the dark forest understory.

Parasitic plants have provided compelling evidence for the phenomenon of exchange of genes not by inheritance, but by what is known as horizontal gene transfer – the exchange of genes between host and parasite, probably made easier by their close physical association. In every cell of a plant body there are three genomes: the one composed of several chromosomes in the nucleus, a circular one in the mitochondrion and a smaller circular one in plastids (like chloroplasts). Both the mitochondrion and the chloroplast are really bacteria that over the course of evolution have become part of the plant cell, losing their ability to live freely. When scientists looked at expressed genes (those serving a function in the organism) found in the *Rafflesia* nucleus, they found that many of them were not from the parasite but instead were from the host – they matched sequences from the grape family. *Rafflesia* has the naturally occurring equivalent of GMO – genes from another species as part of its core functioning genetic system. When the mitochondrion was examined, the numbers jumped even higher – a third or more of the genes in the mitochondrial genome of *Rafflesia* were of *Tetrastigma* origin. These doing this work did loads of tests to ensure they were not seeing the results of contamination; this transfer of genes horizontally from host to parasite was real. These researchers suggested that the incorporation of genes from the host into the parasite's DNA was perhaps a way for the parasite to evade detection by the host; *Rafflesia* could be practicing a sort of genomic deception. After all, the host is making food for two, and that certainly uses resources that could be put to better use for the host itself.

Robert Brown named these extraordinary plants for an equally extraordinary person. Thomas Stamford Raffles was the son of a ship's captain and became the

Franz Bauer's painting of the rafflesia flower amazed his contemporaries. *Rafflesia* flowers still fuel imaginations today – among other things they are the inspiration of the Pokémon character Vileplume.

architect of the governments of British colonies in Southeast Asia. Sent to what is now Penang Island in Malaysia for the British East India Company, he learned the local language, making him ideal for positions where the British needed someone who could somewhat fit in. The political landscape in Southeast Asia was complex and ever-changing, jockeying between the Dutch and English for control was fierce. Taking advantage of the French conquest of Holland, Raffles was part of the British force that took over the island of Java – he was made the Governor General of the Dutch East Indies, a position he held for only three years, when the island reverted to Dutch control as part of the settlements from the Napoleonic Wars.

His short time in Java resulted in a book that shaped the way the British looked at the cultures of Southeast Asia – Raffles's *History of Java* was written during a short stay in England. He chronicled the history of the region, focusing on the glories of the past, contrasting it with the decline and decay as depicted in the images of ruins of ancient temples, all drawn in the picturesque style. These images were the sort that triggered stock responses in the public, of a glorious past and a squalid present. The *History of Java* established the image of a region that needed to go back to those ideals of governance based on a class system with individual property, just those things the

British had to offer. It was not that Raffles didn't love the region, he only did what we all do, unconsciously see things from a biased viewpoint. His descriptions coloured the way the public saw Southeast Asia and its peoples and reinforced the idea that empire was necessary and right.

His next posting from the British East India Company was in 1818 to Bengkulu in southwestern Sumatra. Here he again became immersed in the culture and forests of the area; his interest in natural history was an annoyance for his employers, they were interested in profit alone. As during his previous stint, the jockeying between the British and Dutch was intense, and the local sultans played one side off against the other with consummate skill. Raffles was convinced that ending Dutch control in the region depended upon establishing a base for trade with China and Japan, something he couldn't do from Bengkulu, which was a bit out of the way.

He found a spot at the very tip of the Malay Peninsula where the Dutch were out of the picture and set up a trading post there with the consent of the local sultan, who Raffles had smuggled in to take over. A treaty was signed that included quite substantial yearly payments to the local rulers, in exchange for which the British East India Company had a position right on the trading route to China. Raffles went back to Bengkulu, leaving another in charge in the new British trading post of Singapore. Back in Bengkulu tragedy struck. He lost three of his children in quick succession, and he was himself not well. Return to England was the answer.

Once back there, he was beset with difficulties but immersed himself in the world of science, helping to found and being elected the first President of the Zoological Society of London. The pettiness of the East India Company resulted in his not receiving a pension and being required to pay back £22,000 (more than £2 million [almost $3 million] today) for losses incurred during his administration. When he died in 1826 his entire estate was taken by the Company to cover this debt. To top it all off, Raffles was refused burial in his local churchyard due to his stance against slavery – the vicar came from a family who had made their fortune trading in human beings.

Stamford Raffles is remembered today for his great love for the natural history of Southeast Asia, including the largest and strangest of all flowers – *Rafflesia*.

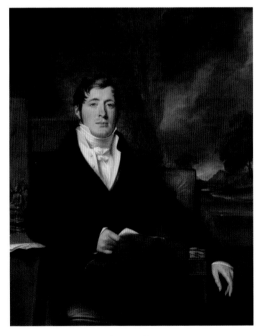

Although Thomas Stamford Raffles (1781–1826) is seen as an epitome of a British colonial administrator, his interest in natural history was not appreciated by his employers.

Sequoiadendron

SEQUOYAH

Family: Cupressaceae, the cypress family
Number of species: 1
Distribution: California

The 'Big Trees' of the Sierra Nevada range in California are among the largest and oldest living things. Reaching to 94 m (more than 300 ft) tall and with diameters of up to 8 m (26 ft) or a bit more, these trees cannot fail to impress anyone who sees them. Individual trees have been dated to more than 3,000 years old, and for many years automobiles were allowed to drive through a tunnel in the Wawona Tree in Yosemite National Park. Standing in a grove of *Sequoiadendron*, surrounded by giants, is an almost mystical experience. Imagine how the frontier scouts led by Joseph Walker, charged with finding an overland route to California over the high Sierra in the early 1830s, must have felt when they crossed the forests where these giants grow. The casual mention of these huge trees in the chronicle of the crossing failed to capture the imagination.

It was not until 1853 that enough material reached British botanists for them to describe the trees botanically. It was collected by Royal Horticultural Society collector William Lobb, who said "This magnificent evergreen tree, from its extraordinary height and large dimensions, may be termed the monarch of the Californian forest." John Lindley, who reported on Lobb's 'discovery', marvelled enthusiastically, "What tree is this! --- of what portentous aspect and almost fabulous antiquity!" Then began the confusion over what to call this marvellous new find. Lindley, in an outrush of patriotic fervour, decided to name the tree *Wellingtonia*, after the British naval hero and previous Prime Minster Arthur Wellesley, 1st Duke of Wellington, who had died the year before:

> "... we think that no one will differ from us in feeling that the most appropriate name to be proposed for this most gigantic tree which has been revealed to us by modern discovery is that of the greatest of modern heroes. WELLINGTON stands as high above his contemporaries as the Californian tree stands above all the surrounding foresters. Let it then bear henceforward the name of WELLINGTONIA GIGANTEA. Emperors and kings and princes have their plants, and we must not forget to place in the highest rank among them our own great warrior."

Sadly for Lindley, the name *Wellingtonia* had already been used several years

The massive Big Trees of the Mariposa Grove of today's Yosemite National Park in the Sierra Nevada were first set aside for "public use, resort and recreation" by President Abraham Lincoln in 1864.

From a Photograph

T. Schrenck 17 N. Hardwicks Edin

SEQUOIA WELLINGTONIA.

MARIPOSA GROVE, SOUTH CALIFORNIA.

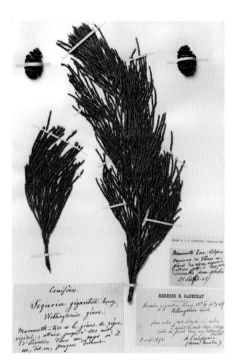

Botanists capture and record plant diversity with herbarium specimens, but the sheer size and magnificence of *Sequoiadendron giganteum* are impossible to capture in two dimensions.

before for a plant from Asia in the flowering plant family Sabiaceae, so the name was already what we call occupied, and not available for use. Even though *Wellingtonia* is now considered a synonym of another genus (*Meliosma* – botanists feel it represents the same biological entity), it cannot be used if we follow the rules of botanical naming. American botanists suggested the genus be named 'Washingtonia', for a great American war hero, but never quite got around to naming the genus and the name *Washingtonia* – honouring George Washington - was used for a genus of palms (see *Washingtonia*). The giant redwoods were then popped in and out of various genera, sometimes with, sometimes without their close relatives, the coast redwoods.

By the 1930s it was clear that the two redwoods were sufficiently distinct to warrant their being recognized as separate genera. But there was no scientific name for the Big Trees – the coast redwoods had a name, *Sequoia sempervirens*, but the magnificent trees of the Sierra Nevada couldn't be called *Sequoia*, they were just too different. Both redwoods were superlative – one the tallest tree and one the largest, but they had different trunk, branch and leaf shapes, and more importantly for botanists, distinct cone morphologies and seed maturation requirements. So, what to call the genus that comprised the Big Trees? The American botanist and conifer specialist John Theodore Buchholz made the practical decision that decided the question:

> "In the selection of an appropriate name for the Big Tree, it is very desirable that no word be selected which is wholly foreign to the names long in use. The name *Sequoiadendron*, which provides adequately for generic distinction, does not wholly discard the long established name, *Sequoia*, and has obvious advantages in catalogues and indices."

So, names can be practical as well as poetic or sycophantic. The name wellingtonia lives on for these trees, though, as the common name for the many individuals cultivated in Britain. In cultivation these trees do not grow as large as those in their home forests of California, but their large stature, beautiful foliage and "awful" image make them much-loved far from home.

The name for the genus *Sequoiadendron* combines the generic name for the coast redwood, *Sequoia*, with the Greek word for tree, so linking these two California trees. The derivation of the name for the Big Trees is therefore obvious, Buchholz made sure of that in his description. But the derivation of the name *Sequoia* is less clear. That genus name was coined by an Austrian botanist Stephen Endlicher in the mid-1840s who thought that the plants called *Taxodium sempervirens* by the British botanist Aylmer Bourke Lambert were distinct enough from *Taxodium* to be made into a new genus. He gave no reasons for his choice of name, as was common at the time.

In the 1860s, a guidebook to the region of Yosemite popularized the idea that Endlicher had intended to honour the extraordinary Native American Sequoyah who invented the Cherokee syllabary in the early 1800s. Sequoyah, who was also known by the English name of George Guest or Gist, was born in the homelands of the Cherokee nation – the Carolinas, Georgia, Tennessee and Alabama - in the late 1700s and raised by his mother in the strongly matrilineal Cherokee tradition. He was monolingual, speaking only Cherokee, an intricate polysynthetic language where single words express complex ideas that would require many words to express in European languages. As a young man Sequoyah was what one might term a jack-of-all-trades – he was a trader, silversmith, blacksmith and soldier – and in his dealings with people of European descent he realized that these settlers had an advantage with their communication via a written language, with which they passed information amongst each other, not relying totally on memory.

He began to develop a plan for a written language for his people, beginning with trying to create a character for every word – but that was too complicated and difficult. As a trader he was able to speak to many people and listen to many conversations; in so doing he analysed the sounds of the language and developed a system where each syllable had a character – he used and adapted some letters of the English alphabet because they were simple to draw, but in the syllabary, they had no relation to their use in English. Sequoyah also developed some new characters of his own. The syllabary was tailored to the sounds unique to the Cherokee language and with its 85 characters represented the full spectrum of sounds used to speak Cherokee, one for each syllable – it was not an imposition of the English alphabet and sounds into the language, it was unique and independent, entirely new.

Sequoyah (*c.* 1770–1843) with a tablet demonstrating the Cherokee syllabary, which is not an alphabet but is rather a set of eighty-five symbols for each syllable in the language.

The idea of an indigenous people independently developing a written language was welcome by some, but others, particularly missionaries, found it problematic because firstly it had been invented by an indigenous person, and secondly this new written language was used by indigenous leaders to record and communicate traditional religious practices. Amongst the Cherokee people the written language was initially viewed with some suspicion; Sequoyah had taught his daughter Ahyoka, and they were accused of witchcraft and put on trial. Their ability to communicate through passing written messages convinced the men from a northern part of the Cherokee nation brought to try them that this invention had value. The story has it that these fierce warriors asked to be taught the system and released the prisoners. In 1825 the syllabary was adopted by the Cherokee Nation and it became widely used extremely rapidly. By 1830 literacy rates amongst the Cherokee peoples were higher than amongst European settlers:

> "The immediate results of what Sequoyah did have no parallel in history.
> Not a schoolhouse was built and not a teacher was hired, but within a few
> months a nation of Indians, call savages by their enemies, rose from a
> condition of savage illiteracy to one of culture, unaided save by one man."

The syllabary facilitated communications between the dispersed Cherokee Nation, who had been forced off their traditional lands by President Andrew Jackson's 'Indian Removal Act' of 1830. Ceding their traditional lands to the east of the Mississippi River to the United States, the Cherokee people were rounded up by the military in the summer of 1838 and removed from their homes to the current capital of the Cherokee Nation in Tahlequah, Oklahoma, but not without cost. Of the 16,000 people who walked the Trail of Tears to Oklahoma some 4,000 died of exposure, starvation, or disease. Following the re-establishment of the Cherokee Nation in Oklahoma, Sequoyah began to contact dispersed bands of Cherokee people who had fled to Mexico so they could be brought back together with those who remained. In so doing he died and is buried in an unmarked grave; but his syllabary lives on and is still in use today.

The idea that Endlicher coined the name *Sequoia* in honour of this extraordinary man comes from his reputation as a linguist and keen admirer of indigenous North American cultures. Endlicher was a fluent speaker of Chinese and many other languages. He certainly knew of Sequoyah, whose story appeared in many German language newspapers of the time, and he collaborated with Steven Peter Du Ponceau, whose work on indigenous American languages included the Cherokee syllabary. But Endlicher didn't say anything about the derivation of the name *Sequoia*. In 2012 it was suggested that instead of being in honour of Sequoyah, the name *Sequoia* was instead a derivation of the Latin verb 'sequor', meaning I follow, and that the honour given the Sequoyah was 'an American myth'. The reasoning was either that *Sequoia* was the follower of genera that had gone extinct or that *Sequoia* came in a sequence of genera distinguished by the number of seeds in each cone scale, thus the generic name for sequence. Subsequent readings of Endlicher's papers have not really solved the mystery but made the derivation from 'sequor' less likely; for a start it would be improper use of Latin, something extremely unlikely for a linguist like Endlicher. No one has yet uncovered hard evidence as to Endlicher's intentions, but it is very right and fitting that two of the most magnificent of American conifers, *Sequoia* and *Sequoiadendron*, honour the inventor of the written syllabary for Cherokee, a hugely important accomplishment that inspired others all over the world to do the same.

Cherokee today is considered a language at risk with only about 2,000 speakers recorded in 2019, but programmes for learning the language are vibrant and evolving. The redwoods too are facing risks in their native habitats; both have declining populations and are considered endangered on the IUCN Red List. The increased incidence and ferocity of fires in the western United States means even forests like those of *Sequoiadendron* that are adapted for fire – with cones that even require fire to open and release their seeds - can be permanently damaged beyond repair. In the Castle Fire of 2020 in the Sierra Nevada more than 10% of *Sequoiadendron* left in the world - some 7,500 to 10,000 trees - were destroyed. A combination of previous policies of fire suppression coupled with hotter and longer dry periods due to climate change means these trees are at the mercy of the heat and ferocity of today's wildfires. Whatever the derivation of their names, the magnificent redwoods of California, both *Sequoia* and *Sequoiadendron*, are reminders of the indigenous peoples of the Americas whose lives were so altered by European settlement and of Sequoyah's unique and lasting contribution to world knowledge. We can only hope they persist into the future in the face of the changing climate the human species has so profoundly altered through our own unsustainable actions.

Despite their immense size and age, *Sequoiadendron* trees are at risk from fires that are hotter and more frequent with climate change. In the devastating 2021 fires individual trees had to be protected with fire blankets.

Sirdavidia

DAVID ATTENBOROUGH

Family: Annonaceae, the soursop family
Number of species: 1
Distribution: Gabon

The only thing my youngest son ever really wanted to watch on television when he was about five or six years old was David Attenborough's multipart series *Life on Earth*. He was gripped by the metaphor of life's history as a clock where humans were a tiny blip just before midnight, the huge diversity of life that we normally don't see, those unicellular organisms to which we are related albeit distantly and, of course the mountain gorillas. Those images of David Attenborough, whispering about the amazingness of an interaction with such extraordinary animals while young gorillas tried to get him to play with them just like human children, did more for showing how much the great apes are like us than textbook descriptions ever could. His obvious joy in interacting with these, some of our closest relatives, was palpable. The inspiration from these and other Attenborough films did not result in my son becoming a professional biologist, but they made a huge difference in framing a small child's outlook on why the rest of life on Earth is so important. Others who did go on to become natural historians, those scientists who document life on Earth, have also been influenced by these films, including the taxonomists who described the extraordinary African plant genus *Sirdavidia* in his honour.

Trained as a zoologist, David Attenborough has been an integral part of natural history on television since its very beginnings, in a way, he defined the genre. Beginning with studio interactions with animals, his passion for life as it is lived in nature led to ground-breaking programmes where plants and animals were filmed where they lived, doing what they do in their everyday lives. First this involved what today would be considered primitive cameras that couldn't get close to the action and sort of simultaneous sound recording that had to be stitched together back in studios, but as technology improved and cameras became more and more sophisticated, Attenborough was able to be part of the action himself, bringing the diversity of nature and his clear joy in it to many viewers. Audiences, first in the UK and later much further afield, could share in the experience – this wasn't necessarily set up scenes in a studio, there were often surprises that came from being in the field, unexpected moments that couldn't have been scripted. From being teams of two men and a camera and sound recorder, the natural history films evolved to become ground-breaking test beds for new photography techniques. Attenborough was also clearly always up for an adventure, romping with mountain gorilla youngsters, being suspended in a cave while bats streamed out, perching on a high branch while talking about the mating displays of birds of paradise in New Guinea, this was nature up close and very personal. A public vote in the UK for favourite 'Attenborough moments' plumped for the shots of him perched on a log observing the extraordinary mimicry skills of the superb lyrebird – in his biography he describes this hilariously; the male lyrebird was attracted using a recording of another bird, but the team had to turn it off before "he [the bird, not Attenborough!] had a nervous breakdown" from not being able to see his rival.

The many honours awarded to David Attenborough attest to his impact on people and institutions all around the world. He has been knighted by Queen Elizabeth II, received awards from museums and governments as well as an Emmy Award; he was voted Britain's most trusted celebrity in a poll run by *Reader's Digest*, recognized as a cultural icon by many, and even had a polar research ship named in his honour. The ultimate impact of making natural history part of everyday life has been immense.

Many plants and animals – extant and extinct – have been named for Attenborough; some even without his knowledge. An extinct plesiosaur originally collected by the great Victorian fossil-hunter Mary Anning was renamed *Attenborosaurus* – the cast of this fossil (the original was destroyed by bombing in

The shape of the flower of *Sirdavidia solannona* inspired the species name – it looks like a member of the Annonaceae, but with the anthers of a *Solanum*.

1940) hangs on the wall in the Natural History Museum, London. When botanists found an unusual new tree in the Annonaceae, the custard apple family, in the rich lowland forests of Gabon in West Africa, they wrote to Attenborough suggesting they wanted to name it in his honour. He replied, "I am truly thrilled that you should have decided to give the new genus of custard apple you have discovered a name based (very ingeniously, if I may say so) on mine. I know very well that such a decision is the greatest compliment that a biologist can pay to another, and I am truly grateful." *Sirdavidia* is the only plant genus named for Attenborough, and it is a plant as extraordinary as he is.

Plant diversity in African tropical forests is lower than that in other tropical regions of the world; the rainforests of both South America and Southeast Asia have significantly more plant species than do those of Africa. This has led to the characterization of Africa as the 'odd man out' because of this lower diversity. Various explanations of this pattern have been suggested over the years, among them poorer knowledge of tropical African rainforests, but recent comparative studies have supported the pattern of lower tropical rainforest diversity on the African continent. Higher extinction rates due to more drastic climatic changes leading to aridification and shrinkage of rainforest areas in the past for the continent do not appear to completely explain this pattern, nor do geological or tectonic factors resulting from the stability of rainforest areas continentally; uplifted areas in Africa tend to be areas of lower rainfall not supporting tropical rainforest, while the opposite is true in the Americas and Southeast Asia. Higher diversification rates, leading to faster accumulation of species diversity, may better explain the differences between tropical rainforest regions – high rates in the Americas and Asia leading to higher diversity, with rates in African rainforest areas remaining constant or lower. But as is usual in our attempts to explain patterns we observe in nature, the underlying causes for these patterns are likely to be complex and interlinked, no single simple explanation like drying due to climate change can completely explain the diversity we see. A complex concatenation of events involving speciation (diversification), extinction, migration, climate, geology and even human impacts remains to be teased out to explain the well-supported 'odd man out' nature of tropical forest diversity in Africa.

But just because on a continental scale African rainforests have fewer species than do those of the Amazon or Borneo, it doesn't mean that African tropical rainforests are not diverse and full of fascinating plants found only there. Gabon, where *Sirdavidia* is endemic, is one of the botanically best-known regions of lowland tropical rainforest in central Africa. The country lies on the west coast of the continent, just where the continent narrows from its broad northern half; 80% of its land area is covered with tropical rainforest and, with neighbouring Cameroon, it forms one of the hotspots of

The understory of tropical forests is incredibly species-rich on all continents, often equalling or surpassing the diversity of canopy trees where so much research is focused.

plant diversity in Africa. Collecting intensity has also been high in the region; Gabon and Cameroon are two of the three (the third is Benin) botanically best-explored countries in tropical Africa. But even here, there continue to be surprises, like *Sirdavidia solannona*.

Far from being found in an under-collected region, *Sirdavidia* was encountered by botanists in the Monts de Cristal National Park, one of the best-collected regions of Gabon, just a few metres away from the road. That this little tree has been overlooked is perhaps due to its habit – the plants are less than 10 cm (4 in) in diameter and tropical forest inventory and collecting efforts focus on trees, and define trees as being large, bigger than *Sirdavidia*. Significant tropical diversity is found however in smaller life forms, in all tropical rainforest regions. So, perhaps these well-known regions are not so well-known after all. Another collection of *Sirdavidia* from central Gabon suggests it might be more widespread; perhaps specimens are lying hidden in herbarium collections, unrecognized because of the unusual flowers of the genus, unusual for the soursop family in any case.

The lineage to which *Sirdavidia* belongs is ancient indeed. Fossils assignable to the Annonaceae are known from the late Cretaceous (*c.* 89 million years ago) and their flowers have often been characterized as 'primitive' (see *Magnolia*). With large, fleshy sepals and petals that often look very much the same, numerous stamens held close together in the centre of the flower, and separate carpels that develop into fleshy structures called monocarps or are sometimes fused together into a massive apple-like fruit – these flowers look like something out of the time of the dinosaurs. Pawpaws, custard apples and cherimoyas are all members of this tropical family. The flowers of *Sirdavidia*, though, are quite outside the norm for the family. Rather than the petals being large and cupped over the stamens and all more or less the same colour, in *Sirdavidia* they

David Attenborough's (b. 1926) enthusiasm for the natural world inspires people of all ages – his sheer enthusiasm when examining a flower and finding out something new is infectious.

are small and swept back, with the highly contrasting stamens forming a cone in the centre of the flower. At first glance, they look like flowers of the nightshade genus, *Solanum*, hence the species name *solannona*. This suggested to the botanists that this might be a flower that was buzz-pollinated – a first for the family, and a first for the magnoliid lineage.

Buzz-pollination is just what it says – buzzing, or sonication, of anthers by bees releases pollen from anthers, which is then transported to another flower by the insect and deposited on the stigma that is usually far-exserted from the anthers. The act of buzzing deposits pollen all over the bee's underside, only some of it is cleaned off and stored in hind-leg pouches, the rest is transported to the next flower they visit. The only reward for pollinators from these flowers is pollen. An important nutritional resource for bees, pollen provides nitrogen for developing larvae, something that is not found in nectar, which provides a concentrated source of carbohydrates, mostly in the form of sugars. Typically buzz-pollinated flowers have anthers that open by terminal pores, tiny holes at the tips of the structures, and highly contrasting anther and petal colours, usually the anthers are bright yellow and look like they are full of pollen, even when they are not. So, *Sirdavidia* fits the bill with respect to the flower colour part, but what about having poricidal anthers? It turns out that some groups of plants that fit the buzz-pollination syndrome have other ways of ensuring their pollen is not accessible except to pollinators who 'properly' vibrate the anthers to release it. Some have anthers that release pollen into a tube formed by the anthers; the structure acts as a single giant pore.

So, maybe this is what is going on with *Sirdavidia* – the cone of anthers, apparently releasing their pollen into the centre, could be grasped by an insect, and buzzed to harvest the pollen for their own use. The problem is, no-one has recorded any insect visitors to *Sirdavidia* – yet. But description of the genus as distinct from any other in the family allows future generations of botanists to investigate the biology of this unusual plant. Or anyone for that matter, as David Attenborough has shown us with his films over the years, many phenomena relating to the day-to-day life of plants and animals in nature can be shown for the first time by photographers who have the patience to wait for something to happen; description of a possibility is an invitation for further work. Of the possibility that *Sirdavidia* is buzz-pollinated, Attenborough mused, "I have filmed carpenter bees extracting pollen from a pink gentian in South Africa, so I am familiar with that technique [buzz pollination] but it would be a particular pleasure to see it being used on *Sirdavidia*…"

Life on Earth is still a work in progress, both through the process of evolution by natural selection and by our increasing understanding of its diversity and variety. We all, from whatever angle we come at it, have a part to play in highlighting its importance to our own futures, as David Attenborough has done throughout his career. The task, if one can call it that, is important, but it is also hugely enjoyable – as Attenborough himself said "I know of no pleasure deeper than that which comes from contemplating the natural world and trying to understand it."

Soejatmia

SOEJATMI DRANSFIELD

Family: Poaceae, the grass family
Number of species: 1
Distribution: Singapore and adjacent Malaysia

What is a grass? Ask that of many people and you might stimulate memories of summer picnics on green lawns, verdant golf courses, or even sometimes the savannas of Africa or South America. Grasses are more than just that though, they are an extraordinary group of flowering plants whose diversity and versatility have long supported human development. Of course, grasses provide us with three of the four main carbohydrate sources in human diets - rice, wheat and maize are all grasses (the fourth carb pillar is the potato, a member of the nightshade family). But grasses are used by people for much more than food. Bamboos, of which *Soejatmia* is one, are a type of grass. They diverse in all regions of the tropics worldwide and are used in construction, local industry and for musical instruments, among many other things. Bamboo is one of the great natural resources of the tropics. Bamboos are often not recognized as grasses; they are usually big and woody, not really the sort of thing one might have as a lawn. They are critically important culturally, especially in Asia, where the diversity of large bamboos is extraordinary. A Chinese philosopher is quoted as saying "A diet without meat makes one emaciated. A living environment without bamboo makes one vulgar. I would rather eat no meat than live without bamboo."

Bamboos are tree-like plants, with long, usually hollow stems (called culms by agrostologists, or grass taxonomists) that can be as much as 25 cm (10 in) in diameter. They have larger leaves than typical lawn grasses and complex underground branching systems, or rhizomes. Bamboo stems are prized for their straightness, smoothness and hardness – they do not have true wood, derived from secondary growth. Their strength comes from a composite structure of fibres mixed with hemicellulose and the organic polymer lignin. The structure is dependent upon both chemical bonds between the molecules and atomic attractions known as van der Waals forces that act when molecules are very close together. The combination of these forces give bamboo stems their extraordinary strength – a bamboo stem has a strength-to-weight ratio higher than either steel or concrete.

This is why bamboo stems are the preferred scaffolding material for skyscraper construction in the cities of Southeast Asia. The length and straightness of bamboo stems is due in part to them not being truly woody; when a stem emerges from an underground rhizome, it is the diameter it will always be, there is no real taper. Hollow

Lemongrass, Cymbopogon (to the left) and bamboo (to the right), here painted by an unknown Chinese artist; both plants are important to the peoples of Asia and both are the subject of intense study by Soejatmi Dransfield.

stems are punctuated with solid sections, called nodes – where the leaves grow. The ideal building bamboos have large diameters, thick walls and relatively short spaces between the solid nodes. The hollowness makes the bamboo scaffolding light, but the composite nature of the walls allows it to flex, making it safer for use in enormous skyscrapers. In the mega-cities of Hong Kong, China and southern China, bamboo scaffolds rising up to 100 stories are erected with lightning speed, and often with minimal safety precautions. Workmen on these scaffolds are said to behave "more like daredvil acrobats than construction workers." It takes seven to nine years to train as a bamboo scaffolder, this is not a profession one can just stroll into. When I saw this for the first time in southern China, the buildings looked as if they were encased in fabulous baskets.

Bamboo is one of nature's ultimate sustainable products. Their growth rate is phenomenal, some species can grow as much as 90 cm (35 in) in a day! Most bamboos, however, don't reach growth speeds one can sit and watch, but are still incredibly rapid, with 10-30 cm (4-12 in) a day being normal. Most of this growth takes place at night. The new shoots are covered with specialized leaves called culm leaves that fall off as the stem elongates. Bamboo shoots are an important vegetable in Southeast Asia; they are usually harvested before they emerge from the ground and become fibrous and hard. Bamboos can fix carbon rapidly and efficiently, making them a nature-based solution to one of the pressing issues of today, human-induced climate change. So, given all of this, one might think that bamboos were well-known and well-studied taxonomically.... Not a bit of it.

Grasses in general are notoriously difficult for taxonomists. Their flowers are small and hard to see (see *Agnesia*), but the approximately 2,000 species of bamboos have a reputation for being impossible, even amongst grass-lovers. They all look very much alike vegetatively – there may not be many different body plans to being a bamboo – the characters in the flowers by which species and genera are defined, are subtle and difficult to see at a glance and, to top it all off, bamboos rarely flower. And when they do, flowering in some results in death of the plant. A once-in-a-lifetime event.

Taxonomists who study bamboo have had lifetime journeys with these wonderful plants, combining careful observation of tiny details with extensive field experience – necessary when working with plants that are not easy to collect, nor easy to find in flower. Such a taxonomist is Soejatmi Dransfield – for whom the bamboo genus *Soejatmia* is named; its author dedicated the genus to her saying "I am pleased to name this rare and beautiful bamboo for Dr Soejatmi Dransfield, whose work has contributed much to the systematic knowledge of Malesian bamboos." Her interest in bamboos began in her native Indonesia, where she worked as an assistant in the national herbarium in Bogor. She became interested in grasses because they seemed easy and, being an excellent field botanist, she was often tasked with taking European and American visitors to the field to find the plants they came to Java to collect. Encouraged by these botanists, she dedicated herself to grasses and to being in the field, where bamboos were ubiquitous and poorly known. Receiving a scholarship to Reading University for her doctoral studies, she completed a thesis on *Cymbopogon* – one of whose species is the delicate herb called lemongrass. But the bamboos called. She realized that field work was essential for the study of bamboos, so back to the field she went – influencing many along the way. Plants in the field inspired and enthused Soejatmi, called Jatmi by her friends and colleagues, whether in the rainforests of Southeast Asia or the temperate forests of upstate New York, where I would accompany her and her husband John in search of rare spring wildflowers. I still think of them whenever I see the New England spring flora. Jatmi and John hold the honour of being the only botanical couple both of whom have a genus named for them – he *Dransfieldia*, a palm (the plant family on which he works), and

Soejatmi Dransfield (b. 1939) in the field in Madagascar, collecting the bamboos whose diversity and novelty she is still uncovering and describing.

she *Soejatmia*, a rare and wonderful bamboo from peninsular Malaysia and Singapore.

The bamboo that became *Soejatmia* was first named as a member of the genus *Bambusa*, a sort of catch-all for woody bamboos, by the British botanist Joseph Gamble in the late 1800s. He had never seen a live plant, so didn't know what size it was or what the stems looked like. All he had was a folded flowering branch, not really an adequate representation of the plant in nature. This specimen had been collected in Singapore by Henry Nicolas Ridley, the colonial director of Gardens and Forests at the time, but whose career had begun at what is now the Natural History Museum. Ridley was responsible for the establishment of extensive rubber plantations in the then-British colonies – ironically contributing to the destruction of the flora he so assiduously documented and explored. Ridley collected what was originally named *Bambusa ridleyi* on Bukit

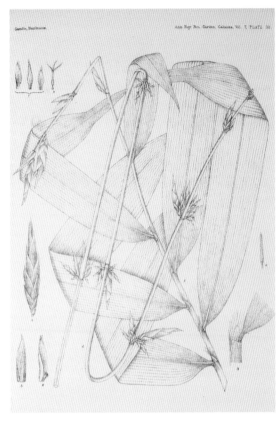

The original illustration of *Bambusa ridleyi* showing details of the flowers and stems; later this species was recognized as different enough from *Bambusa* to name it *Soejatmia*.

Timah, a hill near the geographical centre and the highest natural point in Singapore. It remained poorly known and very rarely collected until Jatmi collected flowering plants in the Malaysian state of Pahang, in the southern part of the Peninsula. These collections led to its recognition as being distinct from *Bambusa*, a result later confirmed using DNA sequence data. But it was that field work that really counted, as Jatmi said later in a paper about Malagasy bamboos, "It is now widely understood that bamboo taxonomic research requires complete specimens, and, for this, fieldwork is essential."

Soejatmia was presented in a paper describing four other new bamboo genera – two of them named for deceased bamboo taxonomists. Jatmi was not supposed to know about the new genus being named for her, John was sworn to secrecy, but as is the case with secrets, it slipped out in casual conversation at a conference. Oops! But noticing the other genera named in the paper Jatmi asked the author – "do you want me to die now then?" I am sure he laughed and said of course not… but it is important to know that new names do not have to honour those who are no longer with us. Those who continue to make an impact on their field are equally worthy of recognition.

When he collected *Soejatmia* in 1890, Nicholas Ridley had to carefully fold the large leaves to fit the size of paper on which his specimen would eventually be mounted.

Soejatmia remains poorly known. It is a small understory bamboo, with stems reaching 5 m (16½ ft) long, and is known only from a few populations; in Singapore it is classified as critically endangered. Because the clustered inflorescences are indeterminate, meaning they continue to produce flowers until the plant reserves are exhausted, it is likely the plants flower and then die (monocarpic), but this is not known for certain. Botanists know so little about bamboo biology beyond a few common species; plants that rarely flower are hard to study, you must be there at just the right time to catch it. Many bamboo specimens are sterile, consisting only of stems and leaves, making it hard to nearly impossible to identify them. Once you do find a bamboo in flower you need that accumulated experience of tiny character variation that can make all the difference in understanding what the plant is.

Based at Kew Gardens, Jatmi worked with and described both genera and species of Southeast Asia bamboos. But colleagues working with the grass flora of Madagascar considered themselves incredibly fortunate that Jatmi Dransfield decided to come to work with them to focus on the bamboos of that island of endemics. Prior to her involvement, knowledge of bamboos in Madagascar was embedded in a narrative of post-colonial study of nature in the global South; plants were described by Europeans or by taxonomists trained in Africa, and pristine forests with 'suitable trees', were valued above other habitats. Bamboos in Madagascar were just bamboos prior to Jatmi's study of them; they were often just placed in already existing genera, – as non-trees they just weren't interesting. She needed to study the bamboos of Madagascar to gain a better understanding of bamboo evolutionary relationships, but their identity was just as important for conservation in Madagascar. She described it as: "What started as a simple exercise in naming a common bamboo has developed into a full-blown reassessment of generic limits." Two of the new genera she described from Madagascar, *Valiha* – named for a musical instrument once made from its culms – and *Canthariostachys* – named for its neat, fan-like inflorescences - are both common large bamboos, impossible to overlook. Despite *Valiha* being a dominant bamboo in many places on the east coast of Madagascar and the most widely used bamboo in the country, Jatmi delayed its description "for many years because I wanted to find and collect a good complete flowering specimen to serve as the type. Unfortunately I have found flowering culms only once…" Both *Valiha* and *Canthariostachys* are also the most important native food sources for the critically endangered greater bamboo lemur, *Prolemur simus*. Knowledge of bamboo distribution compiled by lemur researchers will not only help conservation of these endangered primates, but also aid future study of the unusual bamboos of Madagascar.

Jatmi's work continues to contribute wonderfulness to bamboo taxonomy. Another of the endemic genera from Madagascar draws its name from an endemic Malagasy mammal: "The name *Sokinochloa* was inspired by a remark by one of the guides in Andohahela that it looked like 'sokina', the local name for the greater hedgehog tenrec." That is just what it looks like, a hedgehog balancing on the tip of a bamboo shoot!

It is interesting that many of the great agrostologists are women. Is this because grasses, despite their economic and cultural importance, require patience and dedication to come to know well? These are not the showy flashy members of the flowering plant world, but their diversity and variety are extraordinary, as Jatmi Dransfield has shown in her dedicated study of these un-grass-like grasses.

Woe to anyone who thinks grasses can't have spectacular flowers! The dangling, bright cream-coloured stamens in the flowers of *Canthariostachys* of Madagascar make a striking display.

Strelitzia

SOPHIA CHARLOTTE OF MECKLENBERG-STRELITZ

Family: Strelitziaceae, the traveller's palm family
Number of species: 5
Distribution: southern Africa

Spectacular plants are often chosen as the emblems of places far from their native ranges. Such is the case with the crane-flower or bird-of-paradise plant, *Strelitzia reginae,* that is not only an iconic symbol of its native land, South Africa, where it is the symbol of the South African National Biodiversity Institute and appears on the 50-cent coin, but is also the floral emblem of the city of Los Angeles. *Strelitzia,* when it arrived in Europe, was considered so unusual that when it appeared in a 1790 number of the *Botanical Magazine,* a popular magazine devoted to plant portraits begun by the botanist William Curtis and today known as *Curtis's Botanical Magazine,* two plates were devoted to *Strelitzia* instead of the usual single one, but with the caveat that this really wouldn't happen again. But the excited tones were for an "uncommonly beautiful" plant, whose portrait had been circulated amongst his friends by Sir Joseph Banks – here was the opportunity for everyone to be amazed. Curtis got the flower structure a bit wrong, but the flowers of *Strelitzia* are indeed extraordinary. Curtis waxed lyrical about the flowers themselves "the glowing orange of the Corolla, and fine azure of the Nectary, renders [the flowering stem] truly superb." We now know that the orange structures are the sepals and the arrow-like blue structure that he interpreted as a nectary is two of the three petals that are fused together in a single unit; the tiny third petal is modified into a nectary. The reproductive parts of the flower, the stamens and style, are held within the arrow, with the anthers, full of white pollen, sticking out of the tip. When a bird pollinator, usually a weaverbird or a sunbird, lands on the flower in search of the thick, sweet nectar at the base of the flower, the two sides of the arrow part and the sticky pollen is deposited on the bird's feet, ready to land on the stigma of the next flower visited. All the species of *Strelitzia* have similar flower structure, and all are bird-pollinated, although the colours of the parts vary amongst species.

Strelitzia reginae was first introduced to England through the collections made in South Africa by the plant collector Francis Masson. Masson was the 'replacement' botanist for Joseph Banks on Captain Cook's voyage in HMS *Resolution*; Banks had decided, in a huff, not to go in when the ship could not be altered to accommodate his

The bird-of-paradise 'flower' is really several flowers, opening one after another in a crest-like arrangement; the stiff bird's beak structure holding them is called the spathe.

Strelitzia reginae Banks

The striking combination of bright orange and bright blue attracts pollinating birds to *Strelitzia reginae*, while the stiff bract beneath provides a perch while they probe for nectar.

pack of greyhounds and personal orchestra. Francis Masson was a quiet Scottish under-gardener at the royal gardens at Kew and he became what is described as "Kew's first plant hunter" and an intrepid explorer. His mission was to leave the ship in Cape Town and explore the tip of Africa, then under the control of the Dutch East India Company. Landing in Table Bay near Cape Town in late 1772, he immediately set about exploring, getting lost and generally becoming entranced by the floral riches of South Africa. He travelled by cart, collecting specimens for Banks's herbarium and live plants for the hothouses at Kew. In late 1773 he set off on an expedition to the east, along the coast and inland as far as what is now Port Elizabeth. He was accompanied by a student of Linnaeus's, Carl Thunberg, who was as brash and boastful as Masson was quiet and circumspect; they made a good team though, despite their differing personalities. Along the way they encountered many setbacks – Thunberg's horse fell into a hippopotamus wallow and he nearly drowned; their supply wagon broke down and, to top it all off, they got lost on the way back – twice. But the plants they found were worth it; it was on this expedition that *Strelitzia reginae* was collected, probably somewhere near Port Elizabeth. Once the plant flowered in several of the private gardens in which it had been distributed – one in Essex, one in Banks's own garden in Spring Grove House near Kew – it was a sensation. It was first illustrated in 1777, but it was not until 1787 that Banks organized a painting of his plant from the botanical illustrator James Sowerby that he then had engraved and coloured. This engraving he distributed amongst his friends and colleagues with the name *Strelitzia reginae*, unusually making this privately circulated illustration the place of publication of the generic name! Banks was clearly enamoured of the plant. He dedicated the genus to Queen Charlotte, the wife of George III, whose maiden name was Sophia Charlotte of Mecklenberg-Strelitz, and paid tribute to her position as Queen in the species name – *reginae*. Perhaps a bit over-the-top, but Queen Charlotte was herself a keen botanist and great supporter of the gardens at Kew and of Banks's grand plans for them.

Naming genera for royalty was acceptable in the view of Carl Linnaeus, from whose works all botanical naming begins. He wrote "I retain generic names derived from poetry, imagined names of god, names dedicated to kings, and names earned by those who have promoted botany." Nothing about queens, but Queen Charlotte could certainly be said to have been a promoter of botany.

A princess from a relatively minor 'backwards' German duchy in the northern part of Germany, Sophia Charlotte Mecklenberg-Strelitz was married off to King George III of England shortly after he ascended the throne in 1760. Emissaries from the king described her as ".. not a beauty, but she is amiable"; one point in her favour apparently was that she had no interest in politics. So, in 1761, at the age of seventeen she came to a country where she spoke not a word of the language, was married, and became Queen Charlotte. George III was enamoured of the countryside and or rural living – hence his sobriquet "Farmer George" later in life – so the royal family spent much time in their residences on the outskirts of London, in Richmond and Kew. The couple proceeded to have fifteen children, thirteen of whom survived to adulthood, surely making the country houses a bit cramped! Charlotte suffered not only from isolation due to her not initially speaking the language, which she quickly learned although spoke with a strong accent, but also through the active dislike of her mother-in-law who constantly interfered with family life and appointed many of Charlotte's staff.

When her mother-in-law died, the family took possession of the royal residence at Kew and began to develop the gardens on a grand scale. Both the King and Queen had followed the explorations of Joseph Banks with great interest, and George and Banks had many interests in common – both were titled landowners, interested in agriculture and sheep – they were comfortable together in the informal setting of the countryside. Banks soon became an informal advisor on the development of the royal gardens. The Queen was

Francis Masson (1741–1805) was a plant collector extraordinaire – after South Africa, he went to the Caribbean, and eventually, after being captured by pirates on the way, to North America.

also passionate about gardening, both at Buckingham Palace and at her small cottage near Kew. She and her elder daughters had botany lessons from the savants of the day, flower painting lessons from the premier botanical artists, and assembled their own herbaria on family holidays. Their endeavours were sometimes rather obsequiously praised, although here I am not sure to whom the honour is awarded: the plants or the princesses:

> "There is not a plant at the gardens at Kew (which contain all the choicest productions of the habitable globe) but has been either drawn by her gracious Majesty, or some of the Princesses, with a grace and skill which reflect on these personages the highest honour."

Although the couple were hugely admired by the public for their fidelity and were very close, Charlotte was kept in the dark about the first of George's bouts of manic behaviour in 1765, even though she was to rule should he die or become unable to rule. It was popularized in the film *The Madness of King George* as due to the hereditary condition porphyria, where chemical compounds called poryphyrins build up in the body, although others consider his illness to have been an acute mania in a more psychiatric sense. His second and more serious bout of illness in 1788 distressed her

This herbarium sheet of *Strelitzia reginae* at the Natural History Museum, London was collected from a plant that flowered at the Royal Botanic Gardens at Kew, perhaps from that original material sent by Masson.

immensely and she insisted on accompanying him to Kew where he was treated. *Strelitzia reginae* was dedicated to her around this time, of course for her support of botany and gardening in the royal residences, though Banks probably also knew it was Charlotte who he would need to convince to carry on with his grand plans of a royal garden at Kew. Even if that wasn't the intention, the royal gardens at Kew became the envy of the world, all begun through Banks's relationship with the King and Queen.

Strelitzia belongs to one of the eight families of the plant order Zingiberales – the gingers. These plants are all what one might call mega-herbs – large and non-woody with broad, banana-like leaves. Bird-of-paradise plants can grow to be about 2 m (6 ft) tall, making them something that not everyone has space for in the garden. This was surely true when *Strelitzia reginae* was first introduced to Europe – one had to have a large, well-heated greenhouse in order to grow this "truly superb" plant. If *Strelitzia* strikes you as spectacular, the other two genera in the Strelitziaceae are out of this world –

Strelitzia nicolai, the giant white bird-of-paradise, is a truly spectacular plant; it can be up to 6 m (20 ft) tall with leaves that can each be almost 2 m (6 ft) long arranged in a fan.

Phenakospermum of the Guianas and *Ravenala* (the traveller's palm) of Madagascar are tall, tree-like plants with huge heavy inflorescences growing in tropical rainforests – both are mammal pollinated, in the Guianas by bats and in Madagascar by lemurs.

All *Strelitzia* species are bird pollinated, but not all have the orange and blue flowers of *Strelitzia reginae*; other species have white sepals and either blue or white petals. It has been suggested that the change from white to orange colour in the sepals of *Strelitzia reginae* is related to pollination by weaverbirds, or more plausibly to the habitat in which the plants grow. *Strelitzia reginae* grows in open thickets in the Albany Thicket vegetation of coastal South Africa, where orange would be more noticeable than white. The other species, like *Strelitzia nicolai* are forest plants, where white would stand out against the 'monotony' of green. Birds are also attracted to the seeds of strelitzias; these have a brightly coloured, fat-rich structure called an aril that surrounds them. This aril is bright orange, but with a difference. The orange colour is from bilirubins, previously only known to occur in animals; they are a breakdown product of the precursor of haemoglobin, that gives blood its red colour.

When Curtis described the plant for his avid readers in 1790, he said "... it must remain a very scarce and dear plant." It scarcely needs adding, that it is surely a good plant to name for a queen and, better still, now an emblem for the study of biodiversity in South Africa, one of the great biodiversity hotspots of the globe.

Takhtajania

ARMEN LEONOVICH TAKHTAJAN

Family: Winteraceae, the winter-bark family
Number of species: 1
Distribution: Madagascar

It is hard to imagine that only about 70 years ago there was not widespread acceptance that the continents of the Earth were not always where they are today, that they had moved over geological time. Early in the 20th century the German geologist Alfred Wegener had suggested that the continents had once been a single land mass – today what we call Pangaea. He based his theory on the distributions of fossil plants and animals, and on the outlines of today's continents that seemed to fit together like pieces of a puzzle. Others dismissed these ideas as outlandish, in part because there was no mechanism for this drifting apart proposed. But his ideas did explain the extraordinary patterns of distribution of many plant families confined to the southern Hemisphere. These had been present on the ancient continent of Gondwana and had drifted with the continents to where they occur today. Gondwana was composed of those land masses that are today in the southern hemisphere – largely South America, Australia, and Antarctica, but also India and Madagascar. Gondwanan plants include the southern beeches (*Nothofagus*), the proteas and their relatives, and the Winteraceae, to which the endemic Malagasy genus *Takhtajania* belongs.

The surface of the Earth is made up of large plates, whose edges are constantly being pushed by volcanic and tectonic activity from the centre of the planet. Mountain ranges like the Himalaya or Andes arise when plates collide, this is also where volcanic activity is highest. We now accept that plate tectonics has profoundly shaped the distribution not only of the continents but also of the organisms on them. Early in the discussions, scientists like the Venezuelan Leon Croizat advocated vicariance as the primary mechanism accounting for the distribution of plants – they got where they were today as the result of their past distributions on the proto-continents before their breakup. Others, however, advocated long-distance dispersal over oceans as the primary mechanism accounting for these distributions – as is often true in science when evidence is constantly being discovered, debate raged. As DNA sequence data came to be used in looking at plant evolutionary relationships, the ages of most flowering plant lineages were thought to be too young to have vicariance be the main reason for their distributions on the continent, so long-distance dispersal took a front seat.

But as more and more evidence mounted up, it became clear not only that the origins of flowering plants lay further back in geological time than previously thought, but also that the subsequent ages were a complex mixture of events, some vicariant, some by dispersal over short or long distances. *Takhtajania* in fact seems to be the only flowering plant whose distribution in Madagascar can be explained by Gondwanan vicariance, or the splitting up of that ancient supercontinent. The splitting of Gondwana began about 175 million years ago, with a plate comprising Madagascar–Seychelles–India breaking off some 50 million years later. Connections between these plates were still present until about 116 million years ago – right at the time that the lineage that today only consists of *Takhtajania* diverged. *Takhtajania* is sister to the rest of the Winteraceae – if you think of it as a Y-shaped tree, *Takhtajania* is one branch, the rest of the family comprises the other. This doesn't mean *Takhtajania* is ancient or even that it is primitive, it is just the sole representative of a lineage whose distribution can be explained by movement of the Earth's crust far back in geological time.

The unusual nature of this endemic Malagasy tree's distribution was mentioned when it was first described as a species in the largely Australian genus *Bubbia*; from a single specimen that had been collected more than 50 years before. It wasn't even for sure a *Bubbia* – but was described as one because it didn't really fit anywhere else in the Winteraceae. Subsequent studies, still using just the one collection made in 1909, convinced scientists that a *Bubbia* it wasn't, and a new genus was named "in honour of the eminent systematist and phytogeographer of Leningrad," Armen Takhtajan.

Armen Leonovich Takhtajan did not begin his botanical career in Leningrad (or St. Petersburg as it now again known), but he spent more than half of his long life and much of his scientific career there. Of Armenian heritage, Armen Takhtajan was born

The wet forests of northwestern Madagascar, like these in the Tsaratanana massif, are home to many species that are found nowhere else – both plants and animals.

in Shusha in the disputed region of Nagorno-Karabakh, an island within the republic of Azerbaijan that has long been a zone of conflict between Armenia and Azerbaijan – it is now the Republic of Arksakh. He studied in both Tbilisi (Georgia) and Yerevan (Armenia), combining interests in living and fossil plants. In the late 1930s he became the Director of the Institute of Botany at Yerevan University and founded departments combining the sciences of evolution and palaeobotany at the Armenian Academy of Sciences. He was both nationally and internationally recognized; in 1946 he was awarded a medal for 'Valiant labour in the Great Patriotic War' and in 1947 the Supreme Council of the Armenian Soviet Socialist Republic awarded him a certificate of honour for his training of students at the university. But all was not well, these were times of great debate and stress for science in the then Soviet Union. A photograph of Takhtajan and his colleague Andrey Federov taken in 1944 has inscribed on the back "Comrade censor please skip this photograph! These are the Yerevan botanists Fedorov and Takhtajan"; both men stare out with almost defiant looks, standing their scientific ground. In 1948, under the influence of Trofim Lysenko and his colleagues (see *Vavilovia*), the decision was made that "it was intolerable that the most important areas of biological science … are in the hands of the anti-Michurinites, the Weismanists-Mendelists-Morganists." Even though he had joined the Communist Party of the Soviet Union a few years before, Takhtajan was accused of "Mendelism-Weismanism-Morganism" and dismissed from his post as director of the Institute of Botany and from all posts at the university. In the wake of this purge of evolutionary biologists, thesis topics were proposed at the university on "Mistakes of Professor Takhtajan" – Takhtajan joked that he would be the best supervisor for such a topic.

Despite this censure, he was invited to Leningrad a year later where, at the Leningrad State University, he built a new nucleus of evolutionary botany integrating living and fossil plants. He was an early proponent of this integration, where understanding of living plant relationships was founded on a good understanding of the past. As director of the Komarov Botanical Institute he began a series of publications describing the fossil plants of the Soviet Union, while also developing systems for classifying all living flowering plants. His system is the one I first learned when I began to study botany – it was like those developed by others such as Arthur Cronquist or Robert Thorne in the United States, with whom he worked, but differed in its finer divisions of families, which made it more difficult, but also made the

Armen Takhtajan (1910–2009) was commemorated by his native Artsakh (Nagorno-Karabakh) alongside his medal from the Soviet government and of course, a sprig of *Takhtajania*.

characteristics that defined the families clear. With these other botanists he also developed a system of floristic regions – a way of dividing the world into regions based on plant distributions that was like that proposed for animals by Alfred Russel Wallace in the late 19th century. His system was organized around Kingdoms – like the Paleotropical – each with several regions – like the Madagascan region –that were themselves divided into provinces. In this system, the relationship between Madagascar and the Seychelles of the Indian Ocean was clear – both were in the same Takhtajan floristic region.

Takhtajania perrieri is often described as a 'living fossil', but it is very much a plant of today, at risk as forests are altered by human activity and climate and other environmental change.

Although the 'Takhtajan system' for flowering plant classification has now been superseded by analyses based on DNA sequence data, his influence on phylogenetic thinking about plant evolution was profound. His focus on bringing fossil data to bear on the relationships of living plants, linking geology and botany, is today the norm. So, it is fitting indeed that the plant genus named for him – *Takhtajania* – is one that best demonstrates the effect the geological movements of the Earth's crust have had on plant distribution – it is evidence for vicariance.

Throughout most of Armen Takhtajan's life *Takhtajania* was only known from the single collection made in 1909. Ever since its first description botanists had tried to find the plant again but failed. The area where it had been collected – the Manongarivo Massif – was in the mountains of northwest Madagascar, a region of dense forest and steep slopes, not an easy prospect for looking for this plant, a needle in a haystack. But in the early 1990s Malagasy collectors gathered a plant in a forest about 150 km (93 miles) to the southeast; a small tree, it was not identified in the field. But when the specimens were carefully examined once back in the herbarium – wow! *Takhtajania* had been found again! This was big news – so big that it was published in the journal *Nature*, not a place where botanical discoveries are usually presented. Botanists from Madagascar, France and the United States all went into action together, collecting new material from this population that could be used in detailed studies including those that revealed its relationships to the rest of the family. A specimen from these new collections was presented to Armen Takhtajan for the Komarov Botanical Institute in a special celebration for his 87th birthday – what a present to receive.

The interest that surrounded the rediscovery of *Takhtajania* was in part due to its being known from so little evidence before, but also to its status as a living fossil that dated back to the time of the dinosaurs. Winteraceae were long thought to be "primitive" amongst flowering plants due to the lack of vessels in their wood. Plants conduct water through their stems in the xylem – a system of hollow, dead cells connected by small holes providing interconnectivity – collectively these highly specialized cells are called tracheary elements. Water is pulled up through these cells from the roots to the leaves, powered by evapotranspiration through the stomates, thin holes in the leaf undersides. Imagine the pressure this hydraulic column must be under in tall trees, or even in small trees like *Takhtajania*. Long, thin tracheids are formed from single cells and have holes on their sides, these are found in all vascular plants, but vessels – shorter, fatter cells with flat perforated ends – are found only in flowering plants. That is, in all flowering plants except the members of the Winteraceae. Until the rediscovery of *Takhtajania* in the 1990s, its wood structure was a mystery, but the new collections confirmed that it too had no vessels in its wood, only tracheids. To many this confirmed *Takhtajania* as a 'living fossil', a remnant of ancient plants before vessels evolved and declining relicts of a bygone age when vessels were not needed. Evidence as to the relationships of Winteraceae says otherwise. Winteraceae are nested within other lineages of flowering plants, like waterlilies, magnolias and laurels, that do have vessels. It has been suggested that Winteraceae lost vessels – long thought, along with flowers, to be a key innovation for flowering plants – because they are disadvantageous in climates where freezing occurs; vessels do not do well in repeated freeze-thaw events such as those postulated to have been common in the areas where the Winteraceae lineage evolved all those millions of years ago. Vesselless wood may not be a signature of retention of a character, but instead a derived state that gives these plants an advantage in their environments by enabling them to retain leaves and thus photosynthetic ability during frequent freeze-thaw events. So maybe *Takhtajania* is not a 'living fossil' after all, but instead testament to the resilience and adaptability of plants in the face of environmental change.

In the years since the rediscovery of this small, but spectacular understory tree, botanists have gone on to find one other population of *Takhtajania*. It is classified as endangered in the IUCN Red List, along with many of the other unique plants of Madagascar. Almost two-thirds of the endemic trees of Madagascar are threatened with extinction, from over-exploitation, increase in land conversion for agriculture, and increased fire incidence resulting from climate change. It is up to us all to let this amazing plant, whose rediscovery sparked widespread international interest in plant conservation, survive. In the words of Dr Seuss's book about the small, furry activist, the Lorax, who lamented the destruction of whole forests: "Unless someone like you cares a whole awful lot, nothing is going to get better, It's not."

Vavilovia

NIKOLAI VAVILOV

Family: Leguminosae [Fabaceae], the pea family
Number of species: 1
Distribution: Turkey and the Caucasus to Iran

It is tragically ironic that the scientist who devoted his life to ensuring the agricultural success of the Soviet Union starved to death in prison. Most people have heard of the disastrous impact on agriculture wrought by Trofim Lysenko during the years of Joseph Stalin's rule, but fewer know the story of the extraordinary man he supplanted, Nikolai Vavilov.

Nikolai Vavilov was one of Russia's first geneticists and, in the years leading up to the First World War, studied in England at the John Innes Horticultural Institute with William Bateson, the scientist who first used the term genetics to describe the study of heredity. It is hard to believe now, but at that time the idea of Mendelian inheritance was highly controversial. Today all schoolchildren are taught that characteristics, like pink flowers or wrinkled seeds or blue eyes or curly hair, are the result of genes found on chromosomes and that these genes are inherited from parents in alternate forms – what are called dominant and recessive. Mendel's experiments were done with pea plants; it is fitting therefore, that the genus named for Vavilov – Mendel's great proponent – is also a pea.

Studying with Bateson in England, Vavilov absorbed the new science of genetics, and, after the end of the war, put it to use in a novel and world-changing way. In 1918 he was made a professor of agronomy in Saratov, along the Volga to the south of Moscow. His early trips to Iran and the Pamir mountain ranges in what is now

Throughout his eventful life, Nikolai Vavilov (1887–1913) strove to use the latest science to help people, but his international profile put him at odds with men in power.

The brilliant pink pea-flowers of *Vavilovia formosa* truly justify the common name of this high mountain plant – the beautiful vavilovia.

Tajikistan, collecting locally adapted varieties of grains, had seeded the really big idea that he put to use in service of the Revolution. Vavilov learned from local farmers that locally bred crops did better in local conditions, and he determined to use the new science of genetics to improve crops for the newly emerging Soviet Union. Plant breeding as a science emerged in concert with genetics; it is not surprising that much of the early work in genetics was done with plants. Before the realisation that traits were heritable, crops had been improved certainly, but with knowledge of genetics, crosses with local varieties or wild species could be made to introduce particular traits of interest into crops.

The years following the Revolution in 1917 were wracked with famine and strife; the civil war raged around Saratov, disrupting Vavilov's breeding experiments with a variety of crops. He wrote to a friend "The plot at the farm of the agronomy faculty is now safer because of its distance from the soldier's camp. So far. There are not many soldiers, but we expect an increase in the near future, and therefore the sowings will be in danger. Last year the sunflower crop was completely destroyed." Throughout the civil war Vavilov kept research going. His work was praised by the powers-that-be in Moscow and, in 1920, he delivered a lecture that rocketed him to fame and perhaps precipitated his downfall. His research into local varieties and the wild relatives of crop plants led him to develop what he called 'The Law of Homologous Series in Hereditary

Variation' – clunky title for a very simple idea. With his 'law' Vavilov set up a series of simple rules for plant breeders hunting for new traits for crop improvement. He observed that the same features such as leaf size or stem stiffness could be found in the various stages of related species, genera or even families. So, all plant breeders looking for new variation had to do was fill the gaps in one crop species with forms that had features that been observed in other related species. His discovery was hailed as a breakthrough for crop improvement and likened to that of the Periodic Table of the Elements – "The biologists congratulate their Mendeleev."

Vavilov was offered new agricultural farms around Saratov, money for plant-hunting expeditions around the globe, but he had been appointed as director of the Bureau of Plant Industry (or Bureau of Applied Botany) in Petrograd (later Leningrad, now again St. Petersburg), a top job for a plant breeder bent on improving Soviet agriculture. Vavilov soon moved his institute north into former tsarist palaces and carried on with his work. Plant-hunting expeditions continued to not only the far-flung corners of the Soviet Empire, but to Africa, South America and Asia. At the Bureau of Plant Industry Vavilov's interests expanded beyond wheat and rye into all crops – from eggplants to coffee and tomatoes to watermelons. He began the conception of the big idea that has completely changed plant breeding everywhere. His collecting and exploration with a focus on crops and their wild relatives led to the theory of centres of origin of cultivated plants.

During the great famine of 1924 the Central Committee, headed by Vladimir Lenin, sent Vavilov to the United States to negotiate for food aid – Soviet agriculture benefited from his trip to obtain seeds for future cultivation. Vavilov's experience of repeated famine in post-Revolutionary Soviet Union determined him to use science to improve the situation – "The famine to prevent is the next one. And the time to begin is now." Fortunately for him, a scientific approach to improving agriculture resonated with Lenin, and Vavilov became one of the darlings of Soviet science, with the backing to collect and to visit and network with international scientists whose work he admired.

His success in collecting in "rather difficult" Afghanistan earned him medals and awards: a Gold Medal for the expedition, the Lenin Prize; but by 1925 Lenin had died and the reins of power had passed to Stalin. As Vavilov continued to travel outside the Soviet Union visiting plant breeders and agricultural stations around the world to learn from others' experiences and sending his staff on seed-collecting trips all over the world, the situation at home lurched from crisis to crisis, with famine becoming a regular occurrence. In 1929 he was elected to the academies of several European countries and made the president of the All-Union Lenin Academy of Agriculture; his star was apparently on the rise. But in the same year, Stalin's "Great Break with the Past" was announced, seeking to change the trajectory of misery. The search began for someone to blame for 'wrecking' and those of the scientific elite, so supported in the early days of the Revolution, were in the spotlight. Someone had to take the blame for the suffering of the people, and it was not going to be those in power.

Much of Vavilov and his fellow plant breeders collecting was focused on areas of high diversity in the Soviet state, including the southern area known as the Caucasus – the mountains in far southern Russia and today's republics of Georgia, Armenia and Azerbaijan that extend to Iran in the east and Turkey to the west. These mountain ranges are a global biodiversity hotspot, with many species known only from there, and have long been considered a natural barrier or, more accurately, as a crossroads between Europe and Asia. It was here that the tiny pea now known as *Vavilovia* was collected and described from Mount Khinaliq in what is now Azerbaijan – first as a member of another genus, *Orobus*. This small plant with brilliant magenta flowers grows in scree slopes at high elevations and is known from very few populations, although like most plants, it is not formally assessed on the International Union for the Conservation of Nature's (IUCN) Red List. *Vavilovia formosa* (its species name means beautiful in Latin) is closely related to both peas – our common green peas, *Pisum sativum* – and to the grass pea, *Lathyrus sativus*; both are hugely important grain legumes for both human and livestock consumption. The seeds of legumes, or members of the pea family, are key components of human diets due to their nutritional quality. Often known as pulses they are high in protein, and the plants themselves fix nitrogen, thus improving the soil they grow in.

Vavilovia therefore was a plant of interest to plant breeders for its ability to survive in the harsh mountain climates of the Caucasus, and for its perennial nature – common garden peas are annual, lasting only a year. Perennial crops are better for the environment, they can limit soil erosion, increase nutrient retention and in general are thought to contribute to climate change mitigation and adaptation. So, this high mountain pea was certainly a target for Vavilov's collectors and seeds certainly formed part of the collections. In 1990 cultivation was attempted from seeds in St. Petersburg. Today *Vavilovia* is of increasing interest in plant breeding for climate adaptation. International efforts to bring it into cultivation and to assess the conservation status of remaining populations mean the beautiful pea still has much to contribute to world agriculture.

The botanist Andrey Alexandrovich Fedorov, who recognized the high mountain pea as distinct from all other peas and gave it the name *Vavilovia*, was not the collector of the plant himself nor was he part of Vavilov's team of plant breeders. But clearly he held Vavilov in high regard – naming a new genus for a scientist whose profession was being attacked by those in power cannot have been easy. Fedorov first named this little plant in 1939, the same year Vavilov himself undertook a plant collecting expedition to the Caucasus, his penultimate foray into the field.

By 1939, when Vavilov was in the Caucasus, the entire science of genetics was under attack in the Soviet Union. The "barefoot scientist" Trofim Lysenko, who Vavilov had supported and aided earlier in his career, had developed his ideas for

Vavilovia, unlike our cultivated peas that are annuals, is a perennial plant, persisting for many years in the harsh rocky scree soils in the mountain slopes of its native range.

Davis 24,504 , O.Polunin
Pisum formosum (Stev.) Boiss.

Turkey. Prov. Hakkari: Kara Dag,
11,600 ft. Loose scree. Fls deep
pink. 16 Aug. 1954.

POLLEN SAMPLE TAKEN
FOR BM(NH) COLLECTION

Vavilovia formosa
(Stev.) Fed.
DET (Vicieae)

improving agriculture that were based on the inheritance of acquired characteristics – he promised five- to ten-fold increases in wheat yields through a process of exposing seeds to cold, and Stalin bought it. Lysenko was elevated to the directorship of the All-Union Lenin Academy of Agriculture, supplanting Vavilov, and publication of plant breeding studies involving genetics was prevented. Vavilov continued to defend the science of genetics and its contribution to plant breeding and thus to the prevention of hunger and famines, but to no avail, he was blocked at every turn. A public confrontation between the two sides turned on the acceptance of Marxist theory and its application to agriculture; science was no longer part of the answer. Vavilov was identified as a 'wrecker' and his days were numbered. But he never surrendered – the titles of one of his planned papers for 1940 showed he was still trying to maintain a role – 'Chromosome Theory from the Point of View of Dialectical Materialism'.

But to no avail. In the late summer of 1940 Lysenko sent Vavilov on a collecting trip to the Ukraine, ostensibly to assess the agricultural potential of the newly annexed territory. Here he was arrested by NKVD agents after a fruitful and certainly enjoyable day of collecting rare grasses and herbs. He was whisked to Moscow to undergo interrogation in the inner circle of the NKVD prison system, not a good place to be. For the next year Vavilov was interrogated for what has been calculated to be approximately 17,000 hours; he was accused and found guilty of anti-Soviet activity and of sabotage of Soviet agriculture – both of which he denied. He was sentenced to be shot, but the sentence was commuted to twenty years hard labour. When all political prisoners were evacuated from Moscow ahead of the advancing German army Vavilov was transferred to prison in Saratov, ironically where his career in agriculture had begun. There, in the years of hardship and starvation for all citizens of his country, he died of malnutrition – the man who had tried to save Soviet agriculture for the good of all its people.

When the Stalinist regime ended and the disastrous reliance on Lysenko's unscientific programme was abandoned, Vavilov was 'rehabilitated' in his own country and in 1987 his centenary of his birth was publicly celebrated. The institute he led for so many years was renamed in his honour as the Vavilov Institute of Plant Industry of the Russian Academy of Agricultural Science. Today the institute is a global centre for crop plant research, something of which Nikolai Vavilov would be justly proud. That plant breeding itself has taken a front seat in the quest to adapt agriculture to the threats of climate and environmental change, building on Vavilov's ideas of centres of crop plant diversity and the importance of crop wild relatives, is another of the legacies of this great man, for whom the beautiful pea *Vavilovia* was so aptly named.

Vickia

VICKI FUNK

Family: Compositae [Asteraceae], the daisy family
Number of species: 1, probably extinct
Distribution: southern Brazil

One of the great tragedies of the time we live in today is that some species and even genera are recognized as distinct and described when they are probably already extinct. Such is the case for the genus *Vickia*. A small shrub from the area around São Paulo, Brazil, *Vickia* was first described in the 19th century by the famed synantherologist Christian Friedrich Lessing as *Gochnatia rotundifolia*. And there it stayed, poorly known and rarely collected, until the 21st century and the interest of present-day synantherologists in unravelling the evolutionary relationships in the huge, fascinating and species-rich daisy family, or the composites, so called for their compound 'flowers' made up of many small individual flowers grouped together in a head. So, what is a synantherologist you might ask. Well, put simply, it is a person who studies the composite or daisy family. The word is derived from the French word synanthérées, and it refers to the fused anthers characteristic of the family – look closely at a sunflower and you can see that the anthers in each of the tiny disk flowers in the centre are fused at their edges into a cylindrical ring, a synanther or joined-up anther. This tube of anthers is where the pollen – the structure that carries the male gametes – of the flower is held and when the anthers open and release the pollen, ready for a pollinator to carry it to another flower, it is held inside the tube. As the style and stigma grow upwards through the tube, the pollen is held on specialized hairs beneath the closed stigma and pushed out, ready for pick-up by a foraging bee or fly or another pollinator. Later the stigma

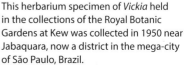

This herbarium specimen of *Vickia* held in the collections of the Royal Botanic Gardens at Kew was collected in 1950 near Jabaquara, now a district in the mega-city of São Paulo, Brazil.

matures and is ready to accept pollen from another plant, ensuring the outcrossing that maintains genetic variation and mixing. But if the flower doesn't get visited by a pollinator with enough pollen to fertilise the ovule, the stigma curls down to touch the remaining pollen from its own flower, ensuring seed production. All this in tiny little flowers – this is how sunflower seeds happen!

The composites are a huge plant family. With some 32,000 species, they rival the orchids for sheer variety. The community of biologists studying the composites – those synantherologists – have worked closely together over the past few decades to understand the evolutionary relationships of the many genera and species of the family. Vicki Funk, for whom the genus *Vickia* is named, was an inspirational force in bringing together knowledge about the family, culminating in a weighty tome, published in 2009. The work involved intensive study of plants in the field, morphological analysis and DNA sequencing – resulting in a picture of the lineages of these plants and their relationships in time and space. Most of composite diversity is found in the lineages we would recognise as asters, sunflowers and their relatives that have explosively diversified in many regions of the globe. Most closely to this hugely diverse set of lineages, that together represent 96% of the species diversity of the family, is the group called subfamily Gochnatioideae, all found in the Americas. When botanists began to look at the subfamily in detail, they realized that the genera recognized were not monophyletic – they contained species that were not each other's closest relatives; *Gochnatia* in particular was a mess. Using a combination of DNA sequence data and morphology – physical features of the plants, like shape of style branches, appendages on anthers and numbers of bristles in the flowers – they split *Gochnatia* up into groups of species that made sense and could be recognized. Sometimes this fractionation of genera is called 'splitting' – taxonomists are often characterized as 'splitters' or 'lumpers' .

Gochnatia rotundifolia was mysterious. Together with another strange species of *Gochnatia* it "lacks any known characters that would allow us to select a genus…. They have only been collected a few times and we were unable to extract usable DNA from existing material." But, although DNA sequence data tells us a lot about evolutionary relationships, it is not infallible. So, when Vicki's colleagues decided to test the uniqueness of *Gochnatia rotundifolia* more closely, they turned to morphology, as Vicki herself would have done. The result was the classification of the species into a new genus – named in honour of Vicki – *Vickia*, still mysterious and rarely collected, but now with a name that honoured one of the great synantherologists of the 20th century.

Vickia is distinguished from the rest of its relatives by combinations of characters of leaves and flowers, especially the round, stiff leaves, which have a distinctive three main veins from their base. This might not seem like much, but its uniqueness in the group and the observation that these patterns are genetically determined convinced botanists that these were "characters that would allow us to select a genus" even though it is a genus of only a single species. They didn't just give up because they didn't have

Cerrado is a shifting mosaic of savanna and sparse forest on low-nutrient and well-drained soils; it is thought to be an ancient vegetation type dating back to the mega-continent of Gondwana.

DNA sequences, they used morphological characters in a cladistic analysis to provide the evidence that supported *Gochnatia rotundifolia* as the sole species of *Vickia*.

Although her great love was for composites – Vicki's office at the Smithsonian was full of sunflower coffee cups and daisy posters – she was also a pivotal part of one of the truly paradigm changing moments in how we understand the world and its relationships, phylogenetics. The modern emergence of phylogeny as a metaphor for our view of life and its inter-relationships is due primarily to an early 20th century German biologist, Willi Hennig. He developed a new way of practicing systematics – the science of classifying the natural world. Hennig's methodology – also called cladistics – involved using characters to group species in a particular way. That wombats and humans share a backbone, four limbs and fur unites them into an inclusive group – this is called a clade, the root of the word cladistics; clades are defined by their special shared characteristics. It does not imply that humans have descended from wombats, or vice versa; it only states that wombats and humans share characters not shared with fish, for example, or insects. The branching diagram depicting these nested sets of shared characteristics – a phylogenetic tree or cladogram – represented a new, more scientific way of looking at nature, through the individual characteristics themselves rather than relying only on expert opinion. Vicki was at the heart of the group of scientists, and one of the first botanists, using Hennigian methods to examine relationships amongst organisms, using her beloved composites. She was often the only or one of very few women in rooms full of arguing

Vicki Funk (1947–2019) travelled far and wide in search of her beloved composites – her work on the biogeography of the Pacific landed her on the summit of Rotui on the island of Moorea.

men; phylogenetic systematics (or cladistics) was the subject of as much scientific controversy as that which surrounded Darwin's original formulation of evolution by natural selection, but with a lot more shouting and swearing.

We know that some things have arisen more than once in the history of life – take wings for example. The wings of butterflies and bats are not structurally the same; rather, the two types of organisms have independently arrived at a 'solution' – flight. The distribution of this character, wings, on a cladogram or phylogeny allows us to make that hypothesis and to use other characters to test it. Each phylogeny then becomes not the truth, but simply a hypothesis about relationships of living things – subject to change and modification as new evidence is found. Both new organisms and new characters can falsify or cause us to reject a phylogenetic hypothesis. Characters that are shared due to common descent – like the fur of mammals or the beaked anthers of the relatives of *Gochnatia* – are called synapomorphies, or more simply, shared derived characters. These can be used to define groups that contain all the relatives of a common ancestor – monophyletic groups. The goal of classification in phylogenetic systematics is the recognition of monophyletic groups, however large or small they might be. In a way, phylogenetics is letting the characters do the talking.

Phylogenetic systematics articulated more than just using characters to examine relationships about the memberships of clades. Hennig also emphasized the importance of what he called 'reciprocal illumination' – always going back to examine the evidence again. Vicki's studies of composite phylogeny were masterpieces of reciprocal illumination. She and her colleagues used the molecular data to create a hypothesis and then tested that with the morphological characters and then looked at the molecular data again; for her, both were important.

This may all seem rather commonplace and obvious now, as we see branching diagrams of coronavirus variants every day, but that we now use phylogeny as the central paradigm for how we see the world is due in large part to those early 20th century pioneers and shouters, even though today some of the methods have moved on from those used by Vicki and her cohort of early cladists.

For Vicki science was a joy, plants were a joy, as were people of all sorts. She was an inspiration to students and colleagues all over the world; when she is remembered it is not only as a scientist, but for her infectious enthusiasm for life, the universe and everything. Her energy in the field collecting was legendary – time getting stuck on Mount Roraima, her excessive cheerfulness in the face of adversity, her dogged determination to go after that one last plant in the ditch, she was an eternal optimist. She also had that sense of humour that made her accessible to anyone, she was "effective in breaking down barriers and puncturing puffery." In today's world of scientific specialism, it is not so common to find a person who has had influence in many different areas of science, but Vicki was one of those. She was that synantherologist, that cladist, but also the woman who championed the importance of collections-based science for natural history museums and its usefulness beyond just classifying and identifying organisms. She went out of her way to help younger female colleagues and was a vocal advocate for women in science – she was an advisor to the American Women's History Initiative, amplifying the contribution of women to the science of botany, leading to the 'Funk List' that is helping to start an avalanche of female scientist inclusion on Wikipedia. She founded learned societies and collaborative networks like the International Biogeography Society and the International Compositae Alliance (TICA) – she was always ready to try a new thing. She shared ideas generously but would argue vociferously if she thought you were wrong. She was a good friend.

It is ironic in a way that the genus named for Vicki, a person so full of life, is probably already extinct. *Vickia* has only been collected 23 times, mostly from the cerrado habitat around the Brazilian megalopolis of São Paulo. The most recent collection of the plant was in 1965 and extensive searches in the region since have all been unsuccessful. When people talk about destruction of habitat in Brazil, most fingers point to the Amazon rainforests – large, green and full of trees, they are the poster child for biodiversity. But other habitats in the region are also biodiversity-rich and full of species that grow nowhere else – the cerrado is one of these. A vast savanna with varying densities of trees, the cerrado ecoregion is found across the southern part of Brazil and neighbouring countries south of the Amazon basin; it is the second largest biome in South America and incredibly biodiverse. Because it doesn't look like a rainforest, it has been considered of little conservation importance; I can remember in a scientific conference someone showing a map of agriculture in Brazil and stating that because it wasn't in the Amazon it was all fine. Not so, the cerrado, with its endemic plants and animals that occur nowhere else on Earth, has been massively impacted by monoculture farming and ranching; it is estimated that only about a fifth of the habitat still exists today. Near São Paulo, the areas of cerrado where *Vickia* once grew are all now under urban sprawl. We might get lucky, and a botanist might find *Vickia* in a remnant patch of cerrado one day, but until then we have only the name, dedicated to the most passionate and enthusiastic of botanists, to remind us that we should cherish and protect the diversity that remains, in all kinds of habitats, with passion and boundless enthusiasm.

Victoria

QUEEN VICTORIA

Family: Nymphaeaceae, the waterlily family
Number of species: 2
Distribution: South America

Much has been written about the giant waterlilies of South America, especially around the time when they were first seen by Europeans. But of course, these magnificent blossoms were long part of the daily life of the peoples of the great South American rivers and were recorded in legends and stories. The Tupi people of the Amazon basin, whose populations were almost completely decimated by disease and mistreatment at the hands of Europeans, tell a tale of forlorn love:

> "The elders of the tribe say that Nauê, the beautiful daughter of the chieftain, fell in love with the moon when she saw its reflection in the water one night. From then on Nauê went to look at the reflection, which, the shaman said, was the prince of the region of Iuaca. After several days the moon disappeared from the lake, and Nauê became very sad and fell sick for a month. One night in her love-sick delirium Nauê saw the moon reflected again in the lake, and to embrace the body of her lover, she threw herself in the water and disappeared. Tupã the god touched by the fate of Nauê transformed her into the most beautiful flower of the lakes…."

The lakes in the story are not lakes as we in the temperate zone might think of them; these are the oxbow lakes left behind as the giant rivers of the Amazon and Paraná change their courses, sometimes connected, sometimes cut off from the main flow, but always full of still water. Both species of *Victoria* are denizens of these still waters. Both inspired their European 'discoverers' to flowery prose in describing their first sight of them and, in England, the flower was likened to the new young queen, Victoria.

Victoria, until recently the longest serving British monarch, was the daughter of the fourth son of George III and was never really expected to rule. A succession of deaths and marriages without children meant that,

The robust veins of *Victoria*, prickly on the undersides, are an architectural network that keep the huge leaves from sinking into the water.

in 1837, she inherited the throne from her uncle William. She was an only child and had grown up highly protected and controlled by her mother; she was very inexperienced in the ways of the world. A month after she turned eighteen, her uncle died – she became Queen Victoria in June 1837. Unmarried still, she had to reside with her mother, something she disliked intensely. It was during this first period of her reign, when she was still an unmarried eighteen-year-old, that her name became associated with the giant waterlily.

Victoria was crowned queen when she was just 18 years old and had until then led a sheltered life; her inexperience meant many sought her favour. Unseemly squabbles between British botanists over the naming of *Victoria*, in private and in public, had much to do with currying favour with the young queen.

Victoria – the plant not the queen! – first came to the attention of British botanists from the letters sent to the Royal Geographical Society by Robert Hermann Schomburgk, a Prussian geographer who was employed in 1835 to explore British Guiana, an amalgamation of the colonies of Essequibo, Berbice and Demerara that had been ceded to the British after the Napoleonic Wars (now the independent nation of Guyana). Schomburgk sent back detailed and highly entertaining accounts of his explorations, with his day-to-day observations. His diary for 1 January 1837 recounts:

> "… Some object on the southern point of the basin attracted my attention; [...] a vegetable wonder! All calamities were forgotten; I felt as a botanist, and felt myself rewarded. A gigantic leaf, from five to six feet in diameter, salver-shaped, with a broad ribbon light green above and a vivid crimson below, rested upon the water; quite in character with the leaf was the luxuriant flower, consisting of many hundred petals, passing in alternate tints from pure white to rose and pink."

Shortly after that, he and his crew were attacked by a herd of white-lipped peccaries, and he had to climb a tree – talk about contrasting experiences! His letters back to the Society in late 1837 included sketches of the flower, causing a sensation in botanical circles. Apparently on the drawing he asked that it might be presented to the new Queen. In a footnote to Schomburgk's account the Editor of the journal waxed sycophantically poetic:

> "To this her majesty has graciously consented, and has also given permission that this flower should be known by the name of "VICTORIA REGIA" Mr Schomburgk will be highly gratified to

learn that his discovery – the most beautiful specimen of the Flora of the western Hemisphere – will henceforward be most appropriately distinguished by the name of our youthful sovereign, herself "the ROSE and expectancy of our state."

It is not recorded what Schomburgk thought, but I am sure he was pleased his find caused a sensation. In the flurry of articles in the British botanical press the plant was referred to as *Victoria regina*, *regia*, *regalis* – in their hurry to publish about this "vegetable wonder" the botanists seemingly lost their ability to spell.

But Schomburgk was not the first explorer to send back an account of the giant waterlily. Eduard Friedrich Poeppig, a German botanist collecting along the Amazon in Brazil, had formally described the same plant as *Euryale amazonica* in 1832, five years before the description of *Victoria*. And the French botanist Alcide D'Orbigny, collecting near Corrientes in Argentina along the Río Paraná, had written about a similar plant, but not named it, in 1835. At a still bend in the river near a tributary called San José he found "une plante que est peut-être l'une des plus belles d'Amerique. [a plant that is probably one of the most beautiful of the Americas]" He described it in detail, also recording the local name in Guarani – yrupẽ, from 'y' meaning water and 'rupẽ' large plate – referring to the resemblance of the leaves to the flat basket covers used by the Guarani people.

As the leaves of *Victoria* expand, the stiff, sharp prickles rip and tear at other plants competing for space, eventually pushing them out of the way and occupying the water surface.

The discovery that *Victoria* was known to more than British botanists, unleashed a set of justificatory articles over the next couple of decades in the British botanical press as to why the name *Victoria regia* had to stand. In these articles misspellings and mistranslations abound. All this has often been attributed to tussles over Empire, but it sounds to me like people wanting to claim rights over something pretty special – a beautiful flower. But botany has rules for naming, and one of these is that the oldest name has priority, it should be used over others proposed after. In 1847 a German botanist, Johann Klotszch, pointed this out and in 1850 James de Carle Sowerby, secretary to the Royal Botanic Society, put it in bald terms:

> ".. the specific name *Amazonica* ought to be retained, or rather, it ought never to have been altered. As for the "permission of Her Majesty," our loyalty need not to be alarmed, for it appears most probable that the "permission" only applied to the name VICTORIA along with the generic name in Sir R. Schomburgk's letter before it was revised, *Regina* being an afterthought."

The rule here, and one to which botanists still adhere, is that the oldest name is the one we use. So, because the giant waterlily was first described by Poeppig as *Euryale amazonica*, it should be called that, unless of course it is recognized as belonging to a different genus than *Euryale*. Comparison of the South American plants with

The up-and-coming botanical illustrator Walter Hood Fitch produced beautiful, delicately hand-coloured plates of the astonishing flowers of *Victoria amazonica* from plants grown in the greenhouses at Syon Park, London.

Euryale ferox, the prickly waterlily or Gorgon plant of India and only species of *Euryale*, revealed many differences, despite the two genera being each other's closest relatives – meaning that the genus name *Victoria* can be used for Schomburgk's finds. But if a species changes genus, the correct species name is still that published first – so *Victoria amazonica* it must be. But *Victoria regia* continued to be used, way into the mid-20th century; now though, we have it right. Botanists have worked hard to make the rules clearer, and it has paid off for the (relative) stability of names today.

D'Orbigny was pretty miffed at having the wondrous plant that he found along the Río Paraná described out from under him; in a paper in 1842 he acerbically noted "en 1837, je vis présenter a l'Académie de Sciences ma plante de la province de Moxos, sous le nom pompeux de *Victoria regia*, donné par M. Lindley. [in 1837 I saw my plant collected in the province of Moxos presented under Lindley's pompous name, *Victoria regia*]." He quickly described the waterlily he had originally collected in the drainage of the Paraná as *Victoria cruziana* – naming it in honour of General Andrés de Santa Cruz, one of the liberators of South America from whom D'Orbigny had received much assistance while in Bolivia. The British botanists persisted in not recognising this as a distinct species, but they were wrong. This species is distinct from that of the Amazon basin; it tolerates cold temperatures better and is the giant waterlily now grown in botanical gardens worldwide outside of the Amazon basin. Both species of *Victoria* have the extraordinary, upturned leaf margins that make the floating leaf into a sort of rimmed plate; the rim on *Victoria cruziana* is a bit taller and the under-surface is purple with a delicate fuzz rather than red and hairless, but the leaves look very similar. As do the flowers, both are large, open at dusk and generate their own heat.

The flowers of *Euryale ferox* of India are nothing like the huge blooms of *Victoria*, but they share extremely prickly flower stalks and calyces.

Victoria plants are anchored to the bottom of the oxbow lakes where they occur, and their flower buds are formed underwater. As the stalks elongate, the buds reach the surface, ready for flowering. The flowers open at dusk, all at once over the surface of the water – a spectacular sight. They emit a strong scent – some say reminiscent of bananas or a mixture of butterscotch and pineapples – and begin to generate heat. Flowers of *Victoria amazonica* measured in their natural habitat can be almost 10°C (50°F) hotter than the air around them! This heat helps to spread the scent, which together with the bright white of the flowers attracts beetles, who feed on starchy structures in the centre of the flower. As the night progresses, the petals gradually begin to change colour to pinkish red and close over the beetles still eating away in the centre of the flower. By dawn the flower is completely closed and stays that way until dusk the next day, when the now purple petals open up, releasing the pollen-covered beetles who fly out and find the white flowers that are just opening, fragrant and ready for pollination. While trapped in the flowers for the day, the beetles eat the starchy appendages in the centre of the flower – a warm stay with food provided.

By the time they open to release the captive beetles, the purple petals are no longer attractive to the insects. Schomburgk had observed beetles in the flowers of *Victoria* in Guyana – he thought they might be species of scarab beetle. Near Manaus, the principal pollinator of *Victoria amazonica* was a new species of scarab; future studies of *Victoria* pollination in the field are sure to turn up more insect novelties as well. Once pollinated, the flowers sink back under the water surface, where the fruit

Victoria cruziana is the most commonly cultivated species in botanical gardens outside of Amazonia. Here in Rio de Janeiro their huge leaves dwarf those of the waterlilies behind.

develops. When the seeds are ripe the fruit wall rots away, and the seeds float to the surface, where they are carried to new suitable habitat by river currents. If the water is too deep, *Victoria* cannot grow, if too shallow it dries out – so the fluctuating water levels of these great South American rivers means that there is always somewhere that is just right for *Victoria* to flourish – so long as the rivers are left to run in their natural way.

Victoria has been an inspiration for many things; it is supposed to have been at the heart of Sir Joseph Paxton's design for the Crystal Palace of the 1851 Great Exhibition in London and more recently has been suggested as the model for the 1914 Glashous building designed by the modernist architect Bruno Taut. That building looked for all the world like the bud of *Victoria amazonica* built entirely of glass. The frenzy that accompanied the giant waterlily's first flower was not confined to the botanical literature, the plant was a star in the popular imagination as well.

Victoria is special. I didn't know the Tupi legend when I first saw it in the waters of a Guyanese river, but it is the image I will keep with me – the maiden who loved the moon, turned to a brilliant white night-blooming flower.

Washingtonia

GEORGE WASHINGTON

Family: Arecaceae, the palm family
Number of species: 2
Distribution: southwestern United States of America to northern Mexico

It is not entirely clear why the 19th century German botanist Hermann Wendland named the elegant palm known to him only from cultivation in honour of the 18th century hero and first President of the United States of America. All he said was "For this plant, previously known as the Brahea or Pritchardia filifera, I propose the generic name Washingtonia, as a reminder of the great American." Perhaps he had in mind John Adams's list of characteristics that had propelled Washington to greatness – his handsome face, tall stature, elegant form and graceful attitudes and movements.

George Washington could never have even seen or heard of the wonderful palm named in his honour – *Washingtonia* grows in the deserts of what is now the southwestern United States and northern Mexico. The first written description of *Washingtonia* came in the mid-19th century, from the military reconnaissance party mapping arid and mountainous lands between the Missouri River and the coast of California. On the 28th of November 1846 Major W. W. Emory wrote:

"... at the end of the creek, several scattered objects were seen, projected against the cliffs, hailed by the Florida campaigners, several of whom were along, as old friends. They were cabbage trees, and marked the *locale* of the spring, and a small patch of grass."

Less than ten years later, a party of railroad surveyors, stopping at what is now Palm Springs, commented that "The surrounding desert and palm tree gave the scene an oriental aspect." Both sets of travellers were seeing *Washingtonia* in its native habitat, but its presence in the botanical literature was not recorded until somewhat later, when trees collected during surveys of the Mexican boundary were identified with some

Palms, with their huge leaves, are tricky to press as herbarium specimens. Here the leaf base has been cut away and spread to show the fibrous edges.

hesitation as "Brahea? dulcis? Mart." – a reference to a similar palm species occurring from Texas to Nicaragua.

Like *Brahea*, *Washingtonia* is a fan palm, so named for its fan-shaped leaves with the leaflets all arising from a central point, unlike some other palms whose leaflets arise all along the leaf axis – think coconut palm leaves. The leaflets are edged and somewhat connected by stringy white fibres, giving the California species its name, *Washingtonia filifera*. They are extremely tall, fast-growing trees, reaching 25 m (82 ft) in height, with graceful clusters of leaves at the top and, in nature, spectacular 'skirts' of dead leaves sheathing the trunks. Like most palms, *Washingtonia* are single-stemmed with the growing point at the tip – this growing point provides us with delectable heart of palm, but its harvest in most cases kills the palm. Palms are called trees, but that is technically incorrect; true trees have a layer of cambium that adds growth, wood (called secondary growth), to the trunk year after

Old leaves of *Washingtonia filifera* form huge hanging skirts that provide safe nesting sites for birds and small mammals but are a fire risk when the plant is cultivated.

year. Palms have no such secondary growth, their solid, strong trunks are not wood but composed entirely of fibres. *Washingtonia* palms are certainly handsome, tall, elegant and graceful.

The exploration of western North America accompanying the wholescale expansion of the United States to the Pacific coast was bringing many new plants, animals and landscapes to the attention of both American and European scientists. Of course, the indigenous peoples living in these regions already knew and used much of this diversity, so it cannot truly be said that the exploring parties 'discovered' the plants. *Washingtonia* fronds were used in shelter construction, the fibres in sandal and basket making, and the fruits eaten. The isolated fan palm oases were used intensively – the dry skirts of dead leaves covering the trunks were regularly burned by local peoples to facilitate fruit collection. These skirts today are cut away from most *Washingtonia* palms in cultivation as a fire precaution in the fire-prone areas of California where it is commonly cultivated.

Washingtonia was brought to Europe by horticulturists intent on providing novelties, probably from seeds collected from cultivated palms in southern California, in the late 19th century. There they were grown as curiosities in botanical gardens – either under glass or in the open in areas where the climate permitted – as *Brahea* or *Pritchardia filifera*. It is these plants that Hermann Wendland – by then one of the foremost palm experts worldwide – saw and realized they were something new to science. Wendland trained in botanical gardens in Germany, Austria and England, and developed a keen interest in the palms. In the 1850s he went to Central America to see them in their native habitats and, by the time he studied *Washingtonia*, was the "most eminent palmographer of the time".

In calling this new palm *Washingtonia*, Wendland didn't realise he was using a name someone had used before – not surprisingly, since Washington was such an American hero. By the strict application of the rules of naming then, this name could not stand. But the other uses of the name *Washingtonia* were fraught with problems; one was not really properly published, the other was suggested but then immediately rejected by its own author, so both these 'washingtonias' were never used (see also *Sequoiadendron*). Just to be on the safe side though, the name *Washingtonia* for the fan palm was conserved against those other names; this means that a proposal was accepted that the name should remain in use for the palm regardless of other, earlier uses of the generic name.

George Washington was not the quintessential self-made American; he was born to wealth and privilege in Virginia, the eldest child of his father's second marriage. He was tall, strong and loved the outdoors. By the age of 16 he was surveying land in the then wild frontier of the Shenandoah valley. His father died young, and George never received a classical education as had his older stepbrothers; his strict, unbendingly critical mother needed his help in running the family estates. His family were Virginia planters, always land rich and cash poor, and for his entire life he struggled with money.

Much against his mother's wishes – she rather peevishly wanted him devoted to her service – Washington joined the army in the early 1750s. Involved in disastrous campaigns along the frontier, his courage and leadership soon vaulted him into the position of commanding the entire Virginia Regiment. But the British army in the North American colonies was not really a meritocracy – colonial officers, like Washington, were paid less than men who came directly from Britain, and on top of that they had to take orders from British officers who were lower in rank. Washington constantly appealed to his superiors to change the situation, but to no avail. These unfairnesses, along with the stranglehold over finance exercised by British suppliers, all contributed to dissatisfaction with the colonial government. Taxes and supplier strangleholds were not good for local business nor the local economy, and the moneyed classes were angry. When a Stamp Act – imposing a tax on all paper – was imposed in 1765, the Virginian Patrick Henry fulminated "Resolved, that the taxation of the people by themselves or by persons chosen by themselves to represent them is the distinguishing characteristic of British freedom." The colonials still thought of

The tiny black fruits of *Washingtonia* are deliciously sweet and date-like; they are produced in huge quantities and were highly prized by the peoples of California before European colonization.

Palm leaves are either pinnate, like those of date palms, with the leaflets all coming off an axis in rows, or palmate, like those of *Washingtonia*, with the leaflets coming from a single place and looking like a fan.

themselves as British, but that was to change. In a meeting of the landed gentry of Fairfax County (Virginia) a group that included Washington agreed that taxation and representation must be inseparable; Washington was appointed the head of a committee to develop responses to increasing British demands. Washington's quiet, listening, evaluating personality soon established him in a leadership position in the group of talkative egomaniacs, like Thomas Jefferson, who comprised the first Continental Congress. He was also a great compromise between the northern and southern colonies – Virginia lay between them and Washington's measured, balanced behaviour was a bridge between the two very different cultures. It is extraordinary that the American Revolution (or War of Independence, depending on who you are) was not a conflict of the masses against the rich but, rather anomalously, was started by the affluent moneyed classes.

By 1775 Washington, as the only one of the group with any real military experience, was the commander-in-chief of the army. Not by looking for the job though, throughout his career George Washington let power come to him. He had to be talked into taking positions of leadership, first the army, then the Presidency. The American rebels were all volunteers, not at all the polished fighting machine of the British who initially picked them off at will and looked down on them as completely hopeless – "Their army is the strangest that ever was collected: old men of 60, boys of 14, and blacks of all ages, and ragged for the most part, compose the motely crew." Washington, however, held them together with inspirational leadership – in the lead-up to the battles following the acceptance of the Declaration of Independence he predicted their place in history "The time is now at hand which must probably determine whether Americans are to be free men or slaves…. The fate of millions will now depend, under God, on the courage … of this army."

The war was long and drawn-out, with many setbacks for the American troops, whose hardship was extreme. In the end, however, they prevailed, not though single great victories, but through grinding attrition of the British forces, who eventually conceded defeat. One virtue of this long conflict was that it gave those in charge time to really think about the form of government they were to establish in their new republic. Many of the things now part of the government of the United States had their genesis in Washington's experiences of trying to raise money for the fighting force; difficulties of raising money from each of the states convinced he and others that there needed to be a central, federal, government.

When the war ended, Washington was lionized by the public of all classes, rich and poor alike. He retreated to his estate on the Potomac River, Mount Vernon, where again he was cajoled and convinced to play his next role in public life – as the first President of the United States. His talents as a public figure were not only his status as an American hero, but also because he controlled his own feelings and opinions, observing well and acting with considerable thought.

George Washington really cared about public opinion and strove throughout his two terms as President to not be flattered by the idolatry that seemed to be his lot, nor to abuse his powers as President. His gift of silence and great sense of self-command made him seem aloof and distant, something his friends denied. His teeth also didn't help, by the time he was President he had lost them all, and dentures in the 18th century were not what they are today.

His second term was fraught with conflict – and established the pattern of American government today, including party politics. The tranquil planter's life Washington craved never was to be his, his retirement from the Presidency to Mount Vernon involved him in continuing money troubles, and the conflicts of his life – over slavery, over appearances, over his role in public life – never abated.

After his death at the end of 1799, George Washington was often portrayed as a stiff, silent figure – the classic "Father of the Nation". The hagiographic, suffocating national piety that followed his demise obscured the truly fascinating character of a man of "died in a manner that befit his life: with grace, dignity, self-possession, and a manifest regard for others." Many of the stories my generation of Americans learned about Washington – his cutting a cherry tree as a child and admitting "I cannot tell a lie", his praying on bended knee before the terrible battles of Valley Forge – are the inventions of a book peddler and itinerant clergyman Parson Mason L. Weems, who "foisted a stiff priggish Washington on the public." In fact, his great rival Thomas Jefferson summed him up best: "…never did nature and fortune combine more perfectly to make a man great."

George Washington's (1732–1799) solemn demeanour in many later portraits may have had something to do with his lack of teeth, rather than any deeper personality traits.

Wuacanthus

WU ZHENGYI

Family: Acanthaceae, the acanthus family
Number of species: 1
Distribution: southwestern China

China harbours extraordinary plant diversity within its borders; with almost 32,000 species of vascular plants (flowering plants, gymnosperms, ferns and lycophytes), half of them endemic and found only in China, the country is the most floristically rich north temperate area of the world. This is in part due to its huge range of habitats, from the deserts of the Mongolian borderlands to the high dry elevations of the Qinghai-Tibet plateau to the tropical rainforests near Vietnam, but that is not all. China has not been heavily glaciated as have been other north temperate landmasses like Europe or North America, and this has left its mark. Tucked at the eastern edge of the vast Qinghai-Tibet plateau and with the Himalayan mountains to the southwest, lie the Hengduan Mountains – the Hengduan Shan – one of the world's thirty-five designated biodiversity hotspots. The region comprises the southern part of Sichuan, the north-western part of Yunnan, and a tiny slice of northern Myanmar. It is characterized by a series of north-south oriented chains of steep mountains rising to more than 5,000 m (16,400 ft) high, interspersed with deep, narrow gorges. The topography seen from above creates a striking picture of the power of geological forces that have shaped our planet. Rivers running south from the Hengduan Mountains form the great watercourses of Southeast Asia – the Brahmaputra, the Irawaddy, the Salween, the Mekong and the Yangtse Rivers all have their sources in or pass through the Hengduan Mountains. Here too live extraordinary animals found nowhere else, like the iconic giant panda and snub-nosed monkeys. And the plant diversity of these mountains is every bit as extraordinary. In an area of about 500,000 square km (193,000 square miles) there have been recorded around 12,000 species of vascular plants - this is more than a third of the total species diversity of the entire country and about the same as the plant diversity of all of Europe.

The origins of this diversity are intriguing – is it due to great age and lack of extinction? Or due to speciation as new habitats and opportunities arose during mountain uplift? Or perhaps colonisation from other areas like the Qinghai-Tibet plateau? Linking earth history with plant diversity patterns can provide explanations as to the major factors involved in such accumulation of diversity. The collision of the tectonic plates – the piece of Gondwana that today makes up India and Laurasia, the supercontinent of the northern hemisphere – that resulted in the crumpling of the

Earth's surface to form the Qinghai-Tibet plateau was more or less complete by the late Miocene period (8–10 million years ago), when mountains surrounding the plateau, such as the Himalaya to the south, had reached their current elevations. The uplift of the Hengduan Mountains by contrast was later – occurring between the late Miocene and the later Pliocene (2–5 million years ago). Analysis of multiple lineages of plants in the Hengduan Mountains has shown that rates of speciation increased drastically during uplift of the mountains, much more so that in the Himalayas or the Qinghai-Tibet plateau. Ecological opportunities for diversification increased with mountain building, with bursts of speciation in the area leading to greatly increased endemic plant diversity.

The north-south orientation of the Hengduan Mountains also allowed the monsoon rains from the south to penetrate far into the valleys – in the east–west oriented mountains of the Himalaya, a significant rain shadow exists. More recent glaciation and the associated changes in climate that occurred in the Pleistocene – these are the glaciations that caused widespread extinction in Europe and North America – were probably not as catastrophic for the plants of the Hengduan Mountains as for other mountain ranges.

That north-south orientation meant that populations could migrate southwards to more suitable ice-free refugia more easily than could those in, for example, the Himalayas, where southwards migration was stopped by huge mountain peaks. During warm interglacial periods, shifts could occur back to higher elevations again open for plant growth. The accumulation of this extraordinary flora was certainly due to many factors working together, and change is still ongoing. Studies on the potential for plant movement due to current climate change have suggested that climate changes will result in range shifts from the Hengduan Mountains into the Qinghai-Tibet plateau, mirroring those shifts in the past. This is not to say that there is no cause for concern, endemic species with narrow ranges may be more negatively impacted – there is still a lot to learn about the flora of the Hengduan Mountains.

That we know so much about the diversity of plant life in these mountains

Wu Zhengyi (1916–2013) travelled widely, here in Edinburgh celebrating the Flora of China project in 2011, enthusiastically promoting Chinese botany and botanists.

was kick-started by the comprehensive study of the entire region undertaken by Chinese botanists between 1973 and 1982. Plant collectors had been to the region, both Europeans and Chinese, but it was the systematic documentation from multidisciplinary Chinese Academy of Sciences expeditions that really consolidated not only new collections and understanding but the rather piecemeal data from earlier exploration. Leading these expeditions to the region was Wu Zhengyi – for whom the genus *Wuacanthus* – endemic to the Hengduan Mountains region in Sichuan and Yunnan – is named.

Wu Zhengyi lived through huge changes in China during his lifetime. Born in 1916 to a family from the town of Yangzhou along the Yangtse river in eastern China, he was educated at home and developed a keen interest in plants from reading classical Chinese texts. While at the National Tsinghua University in Beijing, he became politically active in the 9th and 16th December Movements, student movements demanding that the Chinese government actively resist Japanese aggression. When Japanese forces invaded China and took Beijing in 1937, Wu was on a self-financed botanical collecting trip to Inner Mongolia and had to take a circuitous route to

The slopes of Yulong Xueshan (Jade Dragon Snow Mountain) support a diverse, species-rich flora. Especially prevalent are rhododendrons; the Hengduan Mountains are home to more than two hundred rhododendron species.

return to his hometown. He carried on collecting and botanising throughout the Japanese occupation, moving with others from the university from place to place to the south where the Japanese had not yet reached, often by walking long distances, ending up in Kunming, in Yunnan province. He first encountered the flora of the Hengduan Mountains when he climbed to 4,500 m (almost 15,000 ft) high on the Yulong Xueshan (Jade Dragon Snow Mountain) in the early 1940s; he recorded with pride the more than 2,000 plants they collected on the mountain.

When the occupation ended, Wu joined the movement led by students in Kunming that demanded peace between the different parties in China and advocated democracy, and later joined the forces of the Chinese Communist Party. He served in many important positions across China, but always advocated for the study of the

Wuacanthus was originally described as a species of the genus *Justicia*, and some botanists still consider it belongs there; a name is a hypothesis, subject to test by additional data.

plant diversity of the nation, wherever he was. He led joint teams of Chinese and Soviet botanists on collecting and exploration missions during the rapprochement of the two Communist powers in the 1950s; he made sure the study of China's flora was valued and supported by the government. His 1956 proposal for the establishment of a series of 24 nature reserves across the floristically rich province of Yunnan was accepted; these reserves have made a huge impact on the preservation of biodiversity over the ensuing years of social and environmental change in China.

During the Cultural Revolution (1966-1976) – described by Wu as "The Ten Years of Chaos" – most botanical research was halted, books and papers were not published and, as Wu himself said, "scientific research was frozen and we were sent to undergo criticism and heavy labour." Despite the protracted review of his case by the authorities and his assignment for re-education at a Party establishment, Wu managed to complete four volumes of plant inventories adapted from ancient Chinese text and several other botanical works! Wu emerged from the years of chaos with his reputation rehabilitated and began again to publish and support botanical research in China. It was in these years towards to the end of the Cultural Revolution that concentrated study of the floristic riches of far southwestern China began.

The type of *Wuacanthus* was collected by George Forrest on his eventful travels collecting seeds for keen gardeners fascinated by new plants from southern China.

When Chinese-American relations began to thaw in the late 1970s, Wu was deputy head of the first delegation of Chinese botanists to visit the United States; he travelled to many botanical establishments, making lasting collaborations that resulted in great strides in understanding the diversity of the Chinese flora. The regard in which he was held both by those in power and by the botanical community was reflected in his repeated election to serve in Party Congresses and in his leadership of the compilation of the Chinese language flora of the country.

When the proposal for an English language jointly revised edition of the Chinese flora involving foreign and Chinese authors collaboratively doing field work, research and re-analysis was implemented in the late 1980s, who better to lead from the Chinese side but Wu Zhengyi? The English language *Flora of China* was completed in 2014, leaving a strong legacy of scientific cooperation that continues to this day. This international collaborative effort reflects the core values of this scholarly man, a member of the first generation of botanists to be trained in China. His colleague, Peter Raven, put it best:

"He understood and dealt with the many changes that occurred in China during his life, adapted to them, and was consistently a good citizen as well as an outstanding scientist. … Always ready to encourage others, he fostered the field of Chinese botany and left behind a lasting legacy in the state of its development now."

Wuacanthus was described in his honour in a publication celebrating Wu Zhengyi's life and contribution. The genus has but a single species found in the thickets at low elevations in the dry, deep valleys of the Hengduan Mountains region. It is an unassuming plant, small and scraggly, like many other members of its family, not one of the flashier endemics of the area, like the many species of *Rhododendron*. It was originally described as a member of another genus – *Justicia* – based on early collections of the British botanist George Forrest. The plant was poorly known – the most recent collections were from the 1980s. Rediscovery allowed a Chinese team to study it carefully, using both morphology and DNA sequences – based on their evidence they concluded it was different enough from other plants in the family to warrant its recognition as a new genus.

Others disagree. They feel that more representatives of other genera should have been included in the analyses, particularly those using DNA sequences. They think that *Wuacanthus* is not a "good" genus. Is this a problem? Certainly not. The recognition of taxonomic divisions in life – at all categories – are hypotheses based on the evidence a scientist has available to them. A new species or a new genus is a hypothesis, and like all hypotheses is testable with new or additional evidence. This is what makes taxonomy a science and not just "stamp-collecting" as many have described it. Our knowledge of life on Earth is far from complete, new evidence is being found in the form of new collections and records of species and new techniques are constantly being brought to bear on all kinds of questions. When I began my career the use of DNA sequences to examine evolutionary relationships was in its infancy, it is now the norm. Things change, that is what makes science exciting and so rewarding.

Would Wu Zhengyi have minded that there were differing opinions about the genus named in his honour? Again, certainly not. He was no stranger to having hypotheses overturned – he re-examined the evidence that *Acanthochlamys*, a small herbaceous plant endemic to the high elevation meadows of Hengduan Mountains, should be recognized as its own family, a position held by most Chinese botanists, and said:

"I have the opportunity to reexamine simultaneously the materials available, thus a new and more reasonable morphological interpretation was made in the result to transfer this genus to Velloziaceae (cfr. the family concept adopted by Cronquist 1981 and Dahlgren et al. 1985)."

He looked at the evidence and changed his mind. Decisions as to the distinctness or not of plant genera are our best estimates of the true pattern of evolution using the evidence we have. They are there to be tested, interrogated, and re-examined by new generations of botanists. When things change, it means we have learned something new.

Index

Plant Phylogeny and Bibliography

In the hierarchical classification system used for plants, genera belong to families, families to orders and orders to broader lineages, most of which are given informal names. Tradition is that the person who coined the name for a genus is written out after the genus name, but that family and order names (and above) do not have these names written out. The list below gives all the genera treated in this book with their authors, place of first publication, families, orders and higher lineages. In indicating a higher lineage for flowering plants, I have used the smallest named clade as presented in the 2016 classification system of the Angiosperm Phylogeny Group (APG 2016, An update of the Angiosperm Phylogeny Group classification for the orders and families of flowering plants: APG IV. *Botanical Journal of the Linnean Society*, 181: 1-20). For non-flowering plants I have followed the Angiosperm Phylogeny website (http://www.mobot.org/MOBOT/research/APweb/). Family names in square brackets are alternatives that are permitted by the rules of naming (the *Code*).

General

Bonta, M.M., 1991. *Women in the Field: America's pioneering women naturalists*. Texas A&M University Press, College Station.

Bonta, M.M. (ed.), 1995. *American Women in the Field: writings by pioneering women naturalists*. Texas A&M University Press, College Station.

Heine, H. 1967. 'Ave Ceasar, botanici te salutant'. *Adansonia*, sér. 2, 7: 115-140.

Jarvis, C., 2007. *Order Out of Chaos: Linnaean plant names and their types*. Natural History Museum & Linnean Society of London, London.

Mucina, L. and Rutherford, M.C. (eds.), 2006. *The Vegetation of South Africa, Lesotho, and Swaziland*. Strelitzia 19. SANBI, Pretoria.

Vallejo-Marín, M., 2019. Buzz pollination: studying bee vibrations on flowers. *New Phytologist*, 224: 1068–1074.

Wulf, A., 2009. *The Brother Gardeners: botany, empire and the birth of an obsession*. Alfred A Knopf, New York.

Adansonia Linnaeus, *Systema Naturae* ed. 10, 2: 1144, 1382. 1759 (Malvaceae, Malvales, 'Malvids')

Adanson, M., 1756. *A Voyage to Senegal, the Isle of Gorée and the River Gambia*. [Translated from the French by an English gentleman.] J. Nourse and W. Johnston, London.

Adanson, M., 1763. *Les Familles des Plantes*. 2 vols. Chez Vincent, Paris.

Baum, D.A., 1995. The comparative biology and flora biology of baobabs (*Adansonia* – Bombaceae). *Annals of the Missouri Botanical Garde*, 82: 322–348.

Baum, D.A. et al., 1998. Biogeography and floral evolution of baobabs (Adansonia, Bombaceae) as inferred from multiple data sets. *Systematic Biology*, 47(2): 181–207.

Carteret, X., 2012. Michel Adanson au Sénégal (1749–1754): un grand voyage naturaliste et anthropologique du siècle des luminaires. *Revue d'Histoire des Sciences*, 65: 5–25.

Nicholas, J-P., 1963. Adanson: the man, In: *Adanson: the Bicentennial of Michel Adanson's 'Familles des Plantes'*, Part One. Hunt Monograph Series, Vol. 1. Hunt Botanical Library, Carnegie Institute of Technology, Pittsburgh, pp.1–122.

Patrut, A. et al., 2018. The demise of the largest and oldest African baobabs. *Nature Plants*, 4: 4230426.

Wickens, G.E. and Lowe, P., 2008. *The Baobabs: pachycauls of Africa, Madagascar and Australia*. Springer Science +Business Media B.V., London.

Winsor, M.P., 2004. Setting up milestones: Sneath on Adanson and Mayr on Darwin. In: Williams, D.M. and Forey, P.L. (eds.), *Milestones in Systematics*. Systematics Association Special Vol. 67. CRC Press, Boca Raton, pp.1–17.

Agnesia Zuloaga & Judziewicz, *Novon* 3(3): 306. 1993 (Poaceae [Gramineae], Poales, 'Monocots')

Chase, A., 1922. *First Book of Grasses. The structure of grasses explained for beginners*. MacMillan Press, New York.

Chase, A., 1924–1930. Letters and documents pertaining to Brazilian collecting (various items). Smithsonian Institution Archives, accessed in BHL Collection/Smithsonian Field Books Collection/Women in Natural History, https://biodiversitylibrary.org/creator/2689

Chase, A. and Niles, C.P., 1962. *Index to Grass Species*. 3 vols. G.K. Hall & Co., Boston.

Filgueiras, T.S. et al., 2015. Poaceae. In: *Lista de Espécies da Flora do Brasil*. Jardim Botânico do Rio de Janeiro, available at http://floradobrasil.jbrj.gov.br/jabot/floradobrasil/FB193 [accessed 30 August 2021].

Fosberg, F.R. and Swallen, J.R., 1959. Agnes Chase. *Taxon*, 8(5):145–151.

Grande, J., 2016. Novitates Agrostolgicae, V. Generic mergers in the tribe Olyreae. *Boletin del Centro de Investigaciones Biológicas*, 50(1): 9–43.

Henson, P., 2002. Invading arcadia: women scientists in the field in Latin America, 1900-1950. *The Americas*, 58(4): 577–600.

Hitchcock, A.S. and Chase, A., 1950. *Manual of the Grasses of the United States. 2nd edn*. [USDA Misc. Publ. 200]. Government Printing Office, Washington, D.C.

Zuloaga, F.O. and Judziewicz, E.J., 1993. *Agnesia*, a new genus of Amazonian herbaceous bamboos (Poaceae: Bambusoideae: Olyreae). *Novon*, 3: 3–6–309.

Banksia Linneaus filius, *Supplementum plantarum* 15, 126. 1782, nom. cons. (Proteaceae, Proteales, 'Eudicots')

Cavanagh, T. and Pieroni, M., 2006. *The Dryandras*. Australian Plants Society (SGAP Victoria) Inc., Hawthorn.

George, A., 2014. The case against the transfer of

Dryandra to *Banksia* (Proteaceae). *Annals of the Missouri Botanical Garden*, 100: 32–49.

Hughes, R., 2003. *The Fatal Shore: a history of the transportation of convicts to Australia 1878–1868*. Vintage, London (first published in 1987 by Collins Harvill, London).

Mast, A. and Thiele, K., 2007. The transfer of *Dryandra* R.Br. to *Banksia* L.f. (Proteaceae). *Australian Systematic Botany*, 20: 63–71.

Mast, A. et al., 2005. An assessment of old and new DNA sequence evidence for the paraphyly of *Banksia* with respect to *Dryandra* (Proteaceae). *Australian Systematic Botany*, 18: 75–88.

Musgrave, T., 2020. *The Multifarious Mr. Banks: from Botany Bay to Kew, the natural historian who shaped the world*. Yale University Press, New Haven and London.

O'Brian, P., 1988. *Joseph Banks: a life*. Collins Harvill, London.

Wulf, A., 2009. *The Brother Gardeners: botany, empire and the birth of an obsession*. Alfred A. Knopf, New York.

Bougainvillea Commerson ex Jussieu, *Genera Plantarum [Jussieu]* 91. 1789 (Nyctaginaceae, Caryophyllales, 'Superasterids')

Bautista, M.A.C. et al., 2020. Comparative analysis of complete chloroplast genome sequences of wild and cultivated *Bougainvillea* (Nyctaginaceae). *Plants*, 9(12): 1671, doi.org/10.3390/plants9121671.

Brockington, S. et al., 2009. Phylogeny of the Caryophyllales sensu lato: revisiting hypotheses on pollination biology and perianth differentiation in the core Caryophyllales. *International Journal of Plant Sciences*, 170: 627–643.

Hernández-Lesesma, P. et al., 2015. A taxonomic backbone for the global synthesis of species diversity in the angiosperm order Caryophyllales. *Willdenowia*, 45: 281–384 [accessed on Caryophyllales.org on 14 July 2021].

Lack, H.W., 2012. The discovery, naming and typification of Bougainvillea spectabilis (Nyctaginaceae). *Willdenowia*, 42: 117–126.

Ridley, G., 2010. *The Discovery of Jeanne Baret: a story of science, the high seas, and the first woman to circumnavigate the globe*. Crown Publishers, New York.

Ross, M., 1978. *Bougainville*. Gordon & Cremonesi Publishers, London and New York.

Timoneda, A. et al., 2019. The evolution of betalain biosynthesis in Caryophyllales. *New Phytologist*, 224: 71–85.

Yan, J. and Singh, S., 2019. Migration of Bougainvillea and its domestication: a study. *Journal of Greens and Gardens*, 2: 6–12.

Commelina Linneaus, *Species Plantarum* 40. 1753 (Commelinaceae, Commelinales. 'Commelinids')

Faden, R.B., 1993. The misconstrued and rare species of *Commelina* (Commelinaeae) in the eastern United States. *Annals of the Missouri Botanical Garden*, 80: 208–218.

Hardy, C.R. et al., 2009. Floral organogenesis and the developmental basis for pollinator deception in the Asiatic dayflower, *Commelina communis* (Commelinaceae). *American Journal of Botany*, 96(7): 1236–1244.

Heap, I., 2021. *The international herbicide-resistant weed database*. Available online at http://www.

weedscience.org [accessed 18 October 2021].

Isaac, W-A. et al., 2010. Managing *Commelina* species: prospects and limitations. In: *Herbicides – current research and case studies in use.* Intech Publishers, https://www.intechopen.com/chapters/44759, doi:10.5772/55842.

Kriener, J.M. et al., 2018. Population genomics of herbicide resistance: adaptation via evolutionary rescue. *Annual Review of Plant Biology,* 69: 611–635.

Ram, H.Y.M., 2005. On the English translation of Van Rheede's *Hortus Malabaricus* by K.S. Manilan (2003). *Current Science,* 89(10): 1672–1680.

Ulloa, S.M. and Owen, M.D.K., 2009. Response of the Asiatic dayflower (*Commelina communis*) to glyphosate and alternatives in soybean. *Weed Science,* 57: 74–80.

Ushimari, A. et al., 2007. Colored floral organs influence pollinator behaviour and pollen transfer in *Commelina communis* (Commelinaceae). *American Journal of Botany,* 94: 249–258.

Van Rheede tot Drakenstein H.A., 1689–1703. *Hortus Malabaricus.* 12 vols. J. van Someren & J. van Dyk, Amsterdam.

Wijnands, D.O., 1983. *The Botany of the Commelins.* AA Balkema, Rotterdam.

Darwinia Rudge, *Transactions of the Linnean Society of London* 11: 299. 1816 (Myrtaceae, Myrtales, 'Malvids')

Bentham, G., 1865. Notes on the genera *Darwinia*, Ridge, and *Bartlingia*, Ad. Brongn. *Botanical Journal of the Linnean Society*, 9: 176–182.

Briggs, B.G., 1962. The New South Wales species of *Darwinia*. Contributions from the New South Wales National Herbarium, 3: 129–150.

Briggs, B.G., 1964. The control of interspecific hybridization in *Darwinia*. *Evolution*, 18: 292–303.

Darwin, C., 1858. *On the Origin of Species by Means of Natural Selection.* John Murray, London.

Darwin, E., 1789. *The Botanic Garden II. The loves of the plants.* J. Jackson, Lichfield for J. Johnson, London.

Darwin, E., 1796. *Zöonomia; or The Laws of Organic Life.* J. Johnson, London.

Darwin, E., 1800. *Phytologia: or the philosophy of agriculture and gardening.* J. Johnson, London.

Holmes, R., 2008. *The Age of Wonder: how the romantic generation discovered the beauty and terror of science.* Harper Press, London.

Keighery, G.J., 2009. Six new and rare species of *Darwinia* (Myrtaceae) from Western Australia. *Nuytsia*, 19: 37–52.

King-Hale, D., 1999. *Erasmus Darwin: a life of unparalleled achievement.* Giles de la Mare Publishers, London.

Eastwoodia Brandegee, *Zoe* 4: 397. 1894 (Asteraceae [Compositae], Asterales, 'Asterids')

Anonymous, 1906. The California Academy of Sciences. *Science* n.s., 23(595): 824–827.

Beidleman, R.G., 2009. To California with Jepson's "Phyto-Jogs" in 1913. *Madroño*, 56: 49–56.

Brouillet, L. et al., 2009. Astereae. Chp. 37. In: Funk, V. et al., (eds.) *Systematics, Evolution and Biogeography of the Compositae*, International Association for Plant Taxonomy, Vienna, pp.589–629.

Dakin, S.B., 1954. *The Perennial Adventure: a tribute to Alice Eastwood 1859–1953.* California Academy of Sciences, San Francisco.

Gambill, W.G., 1988. The story of Miss Alice Eastwood. *The Green Thumb*, 45(2): 53–72.

Howell, J.T., 1954. Alice Eastwood, 1859-1953. *Taxon*, 3(4): 98–100.

Keeley, J.E. et al., 2017. Characters in *Arctostaphylos* taxonomy. *Madroño*, 64: 138-153.

Esterhuysenia L. Bolus, *Journal of South African Botany* 33: 308. 1967 (Aizoaceae, Caryophyllales, 'Superasterids')

Hilliard, A.M., 1989. New genera, species and combinations in Scrophulariaceae – Manuleae. *Notes from the Royal Botanic Garden Edinburgh*, 45(3): 481–491.

Klak, C. et al., 2003. A phylogenetic hypothesis for the Aizoaceae (Caryophyllales) based on four plastid DNA regions. *American Journal of Botany*, 90: 1433–1445.

Linder, H.P., 2005. Evolution of diversity: the Cape flora. *Trends in Plant Science*, 10: 536–541.

Lindon, H.L. et al., 2015. Fewer than three percent of land plant species named by women: author gender over 260 years. *Taxon*, 64(2): 209–215.

Linder, H.P. et al., 2003. The high-altitude flora and vegetation of the Cape Floristic Region, South Africa. *Opera Botanica*, 121: 247–261.

Oliver, E.G.H., 2007. Obituary. Elsie Elizabeth Esterhuysen (1912–2006). *Bothalia*, 37: 119–127.

Schnitzler, J. et al., 2011. Causes of plant diversification in the Cape Biodiversity Hotspot of South Africa. *Systematic Biology*, 60: 343–357.

Van Jaarveld, E., 2017. *Esterhuysenia* (Ruschioideae) In: Hartmann, H.E.K. (ed.) Aizoaceae, Vol 2: A-G; In: Uggli, E. and Hartmann, H.E.K. (series eds.), *Illustrated Handbook of Succulent Plants.* Springer Verlag, Berlin., pp.579–585.

Franklinia W. Bartram ex Marshall, *Arbustrum Americanum* 48. 1785. (Theaceae, Ericales, 'Asterids')

Cashin, F.J., 2000. *William Bartram and the American Revolution on the Southern Frontier.* University of South Carolina Press, Columbia SC.

Hoffmann, N.E. and van Horne, J.C. (eds.), 2004. *America's Curious Botanist: a tercentennial reappraisal of John Bartram 1699–1777.* The American Philosophical Society, Philadelphia PA.

Isaacson, V., 2004. *Benjamin Franklin: an American life.* (new edn.) Simon & Schuster, New York NY.

Magee, J., 2007. *The Art and Science of William Bartram.* The University of Pennsylvania Press & Natural History Museum, London, University Park PA.

Gaga Pryer, Fay W. Li & Windham, *Systematic Botany* 37(3): 855. 2012 (Pteridaceae, Polypodiales, 'Ferns and Lycophytes')

Gastony, G.J. and Rollo, D.R., 1998. Cheilanthoid ferns (Pteridaceae; Cheilanthoideae) in the southwestern United States and adjacent Mexico – a molecular phylogenetic reassessment of generic lines. *Aliso*, 17: 131–144.

Haufler, C.H. et al., 2016. Sex and the single gametophyte: revising the homosporous vascular plant life cycle in light of contemporary research. *BioScience*, 66: 928–937.

Hietz, P., 2010. Fern adaptations to xeric environments. In: Mehlreter, K., Walker, L.R. and Sharpe, J.M. (eds.), *Fern Ecology.* Cambridge University Press, Cambridge.

Kao, T-T. et al., 2019. Low-copy nuclear sequence data confirm complex patterns of farina evolution in notholaenid ferns (Pteridaceae). *Molecular Phylogenetics and Evolution*, 138: 139–155.

Li, F-W. et al., 2012. *Gaga*, a new fern genus segregated from *Cheilanthes* (Pteridaceae). *Systematic Botany*, 37(4): 845–860.

Mickel, J.T. and Smith, A.R., 2004. *The pteridophytes of Mexico.* Memoirs of the New York Botanical Vol. 88. New York Botanical Garden, Bronx.

Morran, R.C., 2004. *A Natural History of Ferns.* Timber Press, Portland & Cambridge.

Hernandia Linnaeus, *Species Plantarum* 981. 1753 (Hernandiaceae, Laurales, 'Magnoliids')

Michalak, I. et al., 2010. Trans-Atlantic, trans-Pacific and trans-Indian Ocean dispersal in the small Gondwanan Laurales family Hernandiaceae. *Journal of Biogeography*, 37: 1214–1226.

Varey, S. et al., (eds.), 2000. *Searching for the Secrets of Nature: the life and works of Dr. Francisco Hernández.* Stanford University Press, Stanford, CA.

Varey, S. (ed.), 2000. *The Mexican Treasury: the writings of Dr. Francisco Hernández* (transl. by Chabrán, Chamberlin CL, Varey S). Stanford University Press, Stanford, CA.

Hookeria Smith, *Transactions of the Linnean Society of London* 9: 275. 1808 (Hookeriaceae, Hookeriales, 'Bryophytes')

Allan, M., 1967. *The Hookers of Kew.* Michael Joseph, London.

Eckel, P.W., 2014. Hookeriaceae. In: Flora of North America Editorial Committee (eds.) *Flora of North America north of Mexico, Vol 28: Bryophyta, part 2.* Oxford University Press, Oxford, p.248.

Edwards, D. et al., 2015. Could land-based early photosynthesizing ecosystems have bioengineered the planet in mid-Palaeozoic times? *Palaeontology*, 58: 803–837.

Hooker, J.D., 1902. A sketch of the life and labours of Sir William Jackson Hooker. *Annals of Botany*, 16: ix–ccxx.

Hooker, W.J. and Greville, R.K., 1825. On the genus *Hookeria* of Smith of the order Musci. *Edinburgh Journal of Science*, 2: 221–231.

Hooker, W.J. and Taylor, T., 1818. *Muscologica Britannica; containing the mosses of Great Britain and Ireland.* Longman & Co., London.

Mitchell, R.L. et al., 2021. Cryptogamic ground covers as analogues for early terrestrial biospheres: initiation and evolution of biologically mediated proto-soils. *Geobiology*, 19: 292–306.

Robroek, B.J.M. et al., 2017. Taxonomic and functional turnover in European peat bogs. *Nature Communications*, 8: 1161.

Smith, J.E., 1808. Characters of *Hookeria*, a new genus of mosses, with descriptions of ten species. *Transactions of the Linnean Society of London*, 9: 272–282.

Juanulloa Ruiz & Pavón, *Flora peruviana et chilensis prodromus* 27, t. 4. 1794 (Solanaceae, Solanales, 'Lamiids').

Juan, J. and Ulloa, A., 1748. *Relación histórica del viage a la America Meridional hecho de orden de S. Mag. para medir algunos grados de meridiano terrestre y venir por ellos en conocimiento de la verdadera figura y magnitud de la tierra, con otras observaciones astronomicas y phisicas.* 5 vols. Antonio Marin, Madrid.

Juan, J. and Ulloa, A., 1978. *Discourse and*

political reflections on the kingdoms of Peru: their governments, special regimen of their inhabitants, and abuses which have been introduced into one and another, with special information on why they grew up and how to avoid them [edited and translated from the original Spanish by Trepaske, J.J. and Clement, B.A., Introduction by Tepaske, J.J.]. University of Oklahoma Press, Norman.

Orejuela, A. et al., 2017. Phylogeny of the tribes Juanulloaeae and Solandreae (Solanaceae). *Taxon*, 66(2): 379–392.

Whitaker, A.P., 1935. Antonio de Ulloa. *The Hispanic American Historical Review*, 15(2): 155–194.

Lewisia Pursh, *Flora Americae Septentrionalis (Pursh)* 2: 360. 1813 (Montiaceae, Caryophyllales, 'Superasterids')

Ambrose, S.E., 1996. *Undaunted Courage: Meriwether Lewis, Thomas Jefferson, and the opening of the American West.* Simon & Schuster, New York.

Coues, E., 1898. Notes on Mr. Thomas Meehan's paper on the plants of Lewis and Clarks' expedition across the continent, 1804–1806. *Proceedings of the Academy of Natural Sciences of Philadelphia*, 50: 291–315.

Hall, B., 2003. *I Should be Extremely Happy in your Company: a novel of Lewis and Clark.* Penguin, Harmandsworth.

Hershkovitz, M.A. and Hogan, S.B., 2003. Lewisia. In: Flora of North America Editorial Committee (ed.), *Flora of North America north of Mexico.* Vol. 4: pp.476–485. Oxford University Press, New York.

McCourt, R.M. and Spamer, E.E., 2004. On the paper trail in the Lewis and Clark herbarium. *Bartonia*, 62: 1–24.

Meehan, T., 1898. The plants of Lewis and Clark's expedition across the continent, 1804–1806. *Proceedings of the Academy of Natural Sciences of Philadelphia*, 50: 12–49.

Moulton, G.E. (ed.), 1983–2004. *The Definitive Journals of Lewis and Clark. Vol. 1–13.* University of Nebraska Press, Lincoln [print edition]. Available online https://lewisandclarkjournals.unl.edu/ [accessed 20 August 2021].

Pursh, F., 1813. *Flora Americanae Septentrionalis*, Vols. 1 & 2. White, Cochran & Co., London.

Reveal, J.L. et al., 1999. The Lewis and Clark collections of vascular plants: names, types, and comments. *Proceedings of the Academy of Natural Sciences of Philadelphia*, 149: 1–64.

Linnaea Linnaeus, *Species Plantarum* 631. 1753 (Caprifoliaceae, Dipsacales, 'Campanulids')

Blunt, W., 2001. *The Compleat Naturalist: a life of Linnaeus.* Frances Lincoln, London.

Koerner, L., 1999. *Linnaeus: nature and nation.* Harvard University Press, Cambridge & London.

Lindskog, A., 2020. The north: territory and narrated nature. In: Lindskog, A., Stougaard-Nielsen, J. (eds.), *Introduction to Nordic Culture.* UCL Press, London, pp.23–40.

Linnaeus, C., 1736. *Flora Lapponica.* Solomon Schouten, Amsterdam.

Linnaeus, C., 1737. *Critica Botanica.* C. Wishoff, Leiden. [translated to English by Hort, A. (1938) *The* Critica Botanica *of Linnaeus.* Ray Society, London.]

Linnaeus, C., 1737. *Genera Plantarum.* C. Wishoff, Leiden.

Linnaeus, C., 1811. *Lachesis Lapponica; or a tour in Lapland.* Translated and edited by James Edward Smith. White and Cochrane, London.

Linnaeus, C., 1995. *The Lapland Journey: Iter Lapponicum.* Translated and edited by Peter Graves. Lockharton Press, Edinburgh.

Mari Mut, J.A., 2020. *On the Authorship of* Linnaea *and the Etymology of* Moraea. Editiones Digitales, Aguadilla, http://edicionesdigitales.info/ etimologia/linnaeaandmoraea.pdf.

Scobie, A.R. and Wilcock, C.C., 2009. Limited mate availability decreases reproductive success of fragmented populations of *Linnaea borealis*, a rare, clonal self-incompatible plant. *Annals of Botany*, 103: 835–846.

Magnolia Linnaeus, *Species Plantarum* 535. 1753 (Magnoliaceae, Magnoliales, 'Magnoliids')

Aiello, T., 2003. Pierre Magnol: his life and works. *Magnolia*, 74: 1–10.

Azuma, H. et al., 2001. Molecular phylogeny of the Magnoliaceae: the biogeography of tropical and temperate disjunctions. *American Journal of Botany*, 88: 2275–2285.

Magnol, P., 1676. *Botanicum Monspeliense, sive Plantarum circa Monspelium nascentium index.* F. Carteron, Montpellier & Lyon.

Magnol, P., 1687. *Botanicum Monspeliense, sive Plantarum circa Monspelium nascentium index. Adduntur variarum plantarum descriptiones et icones. Cum appendice quae plantas de novo repertas continet et errata emendat.* D. Pech, Montpellier.

Magnol, P., 1689. *Prodromus historiae generalis plantarum, in quo familiae plantarum per tabulas disponuntur.* G. & H. Pech, Montpellier.

Plumier, C., 1709. *Nova plantarum Americanarum genera.* J. Boudot, Paris.

Sauquet, H. et al., 2017. The ancestral flower of angiosperms and its early diversification. *Nature Communications*, 8: 16047.

Stearn, W.T., 1973. Magnol's *Hortus Monspeliense* and Linnaeus's *Flora Monspeliensis*. In: Geck, E., Pressler, G. (eds.) *Festscrift für Claus Nissen, sum siebzigsten Geburtstag.* Pressler, Weisbaden, pp.613–650.

Tiffney, B., 1985. The Eocene North Atlantic land bridge: its importance in Tertiary and modern phytogeography of the northern Hemisphere. *Journal of the Arnold Arboretum*, 66(2): 243–273.

Wang, Y-B. et al., 2020. Major clades and a revised classification of *Magnolia* and Magnoliaceae based on whole plastid genome sequences via genome skimming. *Journal of Systematics and Evolution*, 58(5): 673–695.

Wen, J. et al., 2010. Timing and modes of evolution of Eastern Asian-North American biogeographic disjunctions in seed plants. In: Long, M., Gu, H., Zhou, Z. (eds.), *Darwin's heritage today: Proceedings of the Darwin 200 Beijing Conference.* Higher Education Press, Beijing, pp.252–269.

Wen, J., 1999. Evolution of eastern Asia and eastern North American disjunct distributions in flowering plants. *Annual Review of Ecology and Systematics*, 30: 421–455.

Megacorax S. González & W.L. Wagner, *Novon* 12(3): 361. 2002 (Onagraceae, Myrtales, 'Rosids')

Bebber, D.P. et al., 2010. Herbaria are a major frontier for species discovery. *Proceedings of the National Academy of Sciences USA*, 107(51): 22169–22171.

Gonzalez Elizondo, M.S. et al., 2002. *Megacorax gracielanus* (Onagraceae), a new genus and species from Durango, Mexico. *Novon*, 12: 360–365.

González-Elizondo, M.S. et al., 2017. Diagnóstico del conocimiento taxonómico y florístico de las plantas vasculares del norte de México. *Botanical Sciences*, 95(4): 760–779.

Levin, R.A. et al., 2003. Family-level relationships of Onagraceae based on chloroplast *rbcL* and *ndhF* data. *American Journal of Botany*, 90: 107–115.

Levin, R.A. et al., 2004. Paraphyly in tribe Onagreae: insights into phylogenetic relationships of Onagraceae based on nuclear and chloroplast sequence data. *Systematic Botany*, 29: 147–164.

National Research Council, 1980. *Research Priorities in Tropical Biology.* The National Academies Press, Washington DC.

Raven, P.H., 1987. *We're Killing our World: the global ecosystem in crisis.* John D. and Catherine T. Macarthur Foundation Occasional Papers, Chicago.

Raven, P.H., 2021. *Driven by Nature: a personal journey from Shanghai to botany to global sustainability.* Missouri Botanical Garden Press, St. Louis.

Rzedowski, J., 1978. *Vegetación de México.* Editorial Limosa, México.

Wallace, A.R., 1911. *The World of Life; a manifestation of creative power, directive mind and ultimate purpose.* Chapman & Hall Ltd., London.

Meriania Swartz, *Flora Indiae occidentalis* 823, t. 15. 1800 (Melastomataceae, Myrtales, 'Malvids')

Dellinger, A.S. et al., 2019. Beyond buzz-pollination – departures from an adaptive plateau lead to new pollination syndromes. *New Phytologist*, 221: 1136–1149.

Erezyilmaz, D.F., 2006. Imperfect eggs and oviform nymphs: a history of ideas about insect metamorphosis. *Integrative & Comparative Biology*, 6: 795–807.

Goldenberg, R. et al., 2020. Taxonomic notes in *Meriania* (Melastomataceae) form the Brazilian Atlantic Forest, including new species, a resurrected one and a new synonym. *Phytotaxa*, 453: 218–232.

Harvey, J., 2006. *Maria Sibilla Merian, The Surinam Album: commentary.* The Folio Society, London.

Heard, K., 2016. *Maria Merian's Butterflies.* Royal Collection Trust, London.

Merian, M.S., 1679–1683. *Der Raupen wunderbare Verwandlung und sonderbare Blumennahrung.* 2 vols. Nuremberg.

Merian, M.S., 1705. *Metamorphosis Insectorum Surinamensium.* G. Valk, Amsterdam.

Michelangeli, F.A. et al., 2015. A revision of *Meriania* (Melastomataceae) in the Greater Antilles with emphasis on the status of the Cuban species. *Brittonia*, 67: 118–137.

Van Delft, M. and Mulder, H. (eds.), 2016. *Maria Sibylla Merian Metamorphosis insectorum Surinamensium/Verandering der Surinaamsche insecten/Transformation of the Surinamese insects.* Lanoo Publishers, Tielt & National Library of the Netherlands, Amsterdam.

Quassia Linnaeus, *Species Plantarum* ed. 2, 1: 553. 1762 (Simaroubaceae, Sapindales, 'Malvids')

Bourdy, G. et al., 2017. Quassia "biopiracy" case and the Nagoya protocol: a researcher's perspective. *Journal of Ethnopharmacology*, 206: 290297.

Clayton, J.W. et al., 2007. Molecular phylogeny of the tree-of-heaven family (Simaroubaceae) based on chloroplast and nuclear markers. *International Journal of Plant Sciences*, 168: 1325–1339.

Linnaeus, C., 1763. *Sistens Lignum Quassiae*. Uppsala [dissertation of Carolus M. Blom, of Smoland].

Odonne, G. et al., 2020 [2021]. Geopolitics of bitterness: deciphering the history and cultural biogeography of *Quassia amara* L. *Journal of Ethnopharmacology*, 267: 113456 [available online 10 Nov 2020].

Price, R., 1979. Kwasimukamba's gambit. *Bijdragen tot de Taal-. Land- en Volkenkunde*, 135(1): 151–169.

Vigneron, M. et al., 2005. Antimalarial remedies in French Guiana: a knowledge attitudes and practices study. *Journal of Ethnopharmacology*, 98: 351–360.

Rafflesia R.Brown ex Thomson bis, *Annals of Philosophy* 16(3): 225. 1820 (Rafflesiaceae, Malpighiales, 'Fabids' or 'COM clade of uncertain placement')

Altjunied, S.M.K., 2005. Sir Thomas Stamford Raffles' discourse on the Malay world: a revisionist perspective. *Sojourn: Journal of Social Issues in Southeast Asia*, 20(1): 1–22.

Barkman, T.J. et al., 2008. Accelerated rates of floral evolution at the upper size limit for flowers. *Current Biology*, 18: 1508–1513.

Beaman, R.S. et al., 1988. Pollination of *Rafflesia* (Rafflesiaceae). *American Journal of Botany*, 75(8): 1148–1162.

Brown, R., 1821. An account of a new genus of plants, named *Rafflesia* (Read June 30, 1820) and Additional observations (Read November 21, 1820). *Transactions of the Linnean Society of London*, 13: 201–234, with 8 plates.

Brown, R., 1844. Description of the female flower and fruit of *Rafflesia Arnoldi*, with remarks on its affinities; and an illustration of the structure of *Hydnora africana*. (Read June 17, 1834). *Transactions of the Linnean Society of London*, 19: 221–247, with 9 plates.

Cai, L. et al., 2021. Deeply altered genome architecture in theendoparasitic flowering plant *Sapria himalayana* Griff. (Rafflesiaceae). *Current Biology*, 31: 1002–1011.

Davis, C.C. et al., 2007. Floral gigantism in Rafflesiaceae. *Science*, 315: 1812.

Mabberley, D.J., 1999. Robert Brown on *Rafflesia*. *Blumea*, 44: 343–350.

Molina, J. et al., 2014. Possible loss of the chloroplast genome in the parasitic flowering plant *Rafflesia lagascae* (Rafflesiaceae). *Molecular Biology and Evolution*, 31(4): 793–803.

Nikolov, L.A. et al., 2013. Developmental origins of the world's largest flowers, Rafflesiaceae. *Proceedings of the National Academy of Sciences, USA*, 110: 18578–18583.

Raffles, S., 1830. *Memoir of the life and public services of Sir Thomas Stamford Raffles, F.R.S. &c.; particularly in the government of Java, 1811–1816; and of Bencoolen and its dependencies, 1817–1824; with details of the commerce and resources of the Eastern archipelago; and selections from his correspondence*. John Murray, London.

Thomson, T. (ed.), 1820. Article XI. Proceedings of the Philosophical Societies. Linnean Society. *Annals of Philosophy*, 16: 225–226.

Tiffin, S., 2009. Java's ruined candis and the British picturesque ideal. *Bulletin of the School of Oriental and African Studies, University of London*, 72(3): 525–558.

Xi, Z. et al., 2012. Horizontal transfer of expressed genes in a parasitic flowering plant. *BMC Genomics*, 13: 227

Xi, Z. et al., 2013. Massive mitochondrial gene transfer in a parasitic flowering plant clade. *PLoS Genetics*, 9(2): e1003265.

Sequoiadendron J.Buchholz, *American Journal of Botany* 26: 536. 1939 (Cupressaceae, Pinales, 'Gymnosperms')

Buchholz, J.D., 1939. The generic segregation of the sequoias. American Journal of Botany, 26(7): 535–538.

Davis, J.B., 1930. The life and work of Sequoyah. *Chronicles of Oklahoma*, 8(2): 149–180 https://web. archive.org/web/20171028175529/http://digital. library.okstate.edu/chronicles/v008/v008p149. html.) [accessed August 2021].

Endlicher, S., 1847. *Synopsis Coniferum*. Scheitlin & Zollikofer, Wien.

Farjon, A. and Schmid, R., 2013. *Sequoia sempervirens*. The IUCN Red List of Threatened Species 2013: e.T34051A2841558, https:// dx.doi.org/10.2305/IUCN.UK.2013-1.RLTS. T34051A2841558.en,]downloaded on 13 August 2021].

Lindley, J., 1853. [no title, *The Gardener's Chronicle*, Saturday December 24, 1853]. *The Gardener's Chronicle and Agricultural Gazette*, 1853: 819–820.

Lowe, G.D., 2012. Endlicher's sequence: the naming of the genus *Sequoia*. *Fremontia*, 40(1&2): 25–35.

Muleady-Mecham, N.E., 2017. Endlicher and *Sequoia*: determination of the etymological origin of the taxon *Sequoia*. *Bulletin of the Southern California Academy of Sciences*, 116(2): 137–146.

Schmid, R. and Farjon, A., 2013. *Sequoiadendron giganteum*. The IUCN Red List of Threatened Species 2013: e.T34023A2840676, https:// dx.doi.org/10.2305/IUCN.UK.2013-1.RLTS. T34023A2840676.en [downloaded on 13 August 2021].

Unseth, P., 2016. The international impact of Sequoyah's Cherokee syllabary. *Written Language & Literacy*, 19(1): 75–93.

Sirdavidia Couvreur & Sauquet, *PhytoKeys* 46: 4. 2015 (Annonaceae, Magnoliales, 'Magnoliids')

Attenborough, D., 2014. *Life on Air*. BBC Books, London.

Couvreur, T.L.P. et al., 2009. Molecular and morphological characterization of a new monotypic genus of Annonaceae, *Mwasumbia*, from Tanzania. *Systematic Botany*, 34: 266–276.

Couvreur, T.L.P., 2014. Odd man out: why are there fewer plant species in African rainforests? *Plant Systematics and Evolution*, 301: 1299–1313.

Couvreur, T.L.P. et al., 2015. *Sirdavidia*, an extraordinary new genus of Annonaceae from Gabon. *PhytoKeys*, 46: 1–19.

De Luca, P.A. and Vallejo-Marín, M., 2013. What's the 'buzz' about? The ecology and evolutionary significance of buzz-pollination. *Current Opinion in Plant Biology*, 16: 429–435.

Linder, H.P., 2014. The evolution of African plant diversity. *Frontiers in Ecology and Evolution*, 38(2): 1–14.

Nunes, C.E.P. et al., 2021. Variation in the natural frequency of stamens in six morphologically diverse, buzz-pollinated heterantherous *Solanum* taxa and its relationship to bee vibrations. *Botanical Journal of the Linnean Society*, 197: 541-553.

Sosef, M.S.M. et al., 2017. Exploring the floristic diversity of tropical Africa. *BMC Biology*, 15: 15.

Soejatmia K.M.Wong, *Kew Bulletin* 48(3): 530. 1993 (Poaceae [Gramineae], Poales, 'Commeliniids')

Biodiversity of Singapore online – Critically Endangered https://singapore.biodiversity.online/ species/P-Angi-000970

Dransfield, S. and Widjaja, E.A. (eds.), 1995. *Plant Resources of South-East Asia. No 7: Bamboos*. Backhuys Publishers, Leiden.

Dransfield, S., 1998. *Valiha* and *Canthariostachys*, two new bamboo genera (Gramineae-Bambusoideae) from Madagascar. *Kew Bulletin*, 53(2): 375–397.

Dransfield, S., 2016. *Sokinochloa*, a new bamboo genus (Poaceae- Bambusoideae) from Madagascar. *Kew Bulletin*, 71: 40.

Gamble, J.S., 1896. *The Bambuseae of British India*. Annals of the Royal Botanic Garden of Calcutta vol. VII. Bengal Secretariat Press, Calcutta.

King, T. et al., 2013. Large-culmed bamboos in Madagascar: distribution and field identification of the primary food sources of the critically endangered Greater Bamboo Lemur *Prolemur simus*. *Primate Conservation*, 2013 (27): 33–53.

Salisbury, E.J., 1957. Henry Nicholas Ridley, 1855–1956. *Biographical Memoirs of Fellows of the Royal Society*, 3: 141–159.

Waters, D., 1998. The craft of the bamboo scaffolder. *Journal of the Hong Kong Branch of the Royal Asiatic Society*, 37: 19–38.

Wong, K.M., 1993. Four new genera of bamboos (Gramineae: Bambusoideae) from Malesia. *Kew Bulletin*, 48: 517–532.

Youssefian, S. and Rahbar, N., 2015. Molecular origin of strength and stiffness in bamboo fibrils. *Scientific Reports*, 5: 11116

Strelitzia Banks, *Strelitzia reginae* [illustration] s.n. 1788 (Strelitziaceae, Zingiberales, 'Commeliniids')

Desmond, R., 2007. *Kew: the history of the Royal Botanic Gardens*. 2nd edn. Royal Botanic Gardens, Kew.

Linder, H.P., 2005. Evolution of diversity: the Cape flora. *Trends in Plant Science*, 10: 536-541.

Mabberley, D.J., 2011. A note on some adulatory botanical plates distributed by Joseph Banks. *Kew Bulletin*, 66(3): 475-477.

Schnitzler, J. et al., 2011. Causes of plant diversification in the Cape Biodiversity Hotspot of South Africa. *Systematic Biology*, 60: 343-357.

Linnaeus, C., 1751. *Philosophia Botanica*. R. Kiesewetter, Stockholm.

Curtis, W., 1790. Strelitzia Reginae: Canna-leaved Strelitzia. *Botanical Magazine*, 3: t. 119, 120.

Cron, G.V. et al., 2012. Phylogenetic relationships and evolution in the Strelitziaceae (Zingiberales). *Systematic Botany*, 37(3): 606-619.

Musgrave, W. et al., 1998. *The Plant Hunters*. Seven Dials, London.

Rentoumi, V. et al., 2017. The acute mania of King George III: a computational linguistic analysis. *PLoS ONE*, 12(3): e0171626.

Takhtajania M.A. Baranova & J.-F. Leroy, *Adansonia* sér. 2, 17(4): 386. 1978 (Winteraceae, Canellales, 'Magnoliids')

Beech, E. et al., 2021. *The Red List of Trees of Madagascar*. Botanic Gardens and Conservation International, Richmond.

Dr Seuss, 1971. *The Lorax*. Random House, New York.

Capuron, R., 1963. Présence a Madagascar d'un nouveau réprensant (*Bubbia perrieri* R. Capuron) de la famille des Wintéracées. *Adansonia* sér. 2, 3(3): 373-378.

Carlquist, S., 2000. Wood and bark anatomy of *Takhtajania* (Winteraceae): phylogenetic and ecological implications. *Annals of the Missouri Botanic Garden*, 87: 317-322.

Doyle, J.A., 2000. Paleobotany, relationships, and geographic history of Winteraceae. *Annals of the Missouri Botanic Garden*, 87: 303-316.

Field, T.S. et al., 2002. Hardly a relict: freezing and the evolution of vesselless woody in Winteraceae. *Evolution*, 56(3): 464-478.

Gabrielyan, I. and Kovar-Eder, J., 2010. In memoriam Armen Leonovich Taktajan. *International Organisation of Palaeobotany Newsletter*, 91: 8-9.

Hallam, A., 1994. *An Outline of Phanerozoic Biogeography*. Oxford Biogeography Series 10. Oxford University Press, Oxford.

Leroy, J-F., 1978. Une sous-famille monotypique de Winteraceae endémique a Madagascar: les Takktajanioideae. *Adansonia* sér. 2, 17(4): 383-395.

Schatz, E., 2000. The rediscovery of a Malagasy endemic: *Takhtajania perrieri* (Winteraceae). *Annals of the Missouri Botanic Garden*, 87: 297-302.

Schatz, G.E., 2001. *Generic Tree Flora of Madagascar*. Royal Botanic Gardens, Kew & Missouri Botanical Garden, St. Louis.

Schatz, G.E. et al., 1998. *Takhtajania perrieri* rediscovered. *Nature*, 391: 133-134.

Thomas, N. et al., 2014. Molecular dating of Winteraceae reveals a complex biogeographical history involving both ancient Gondwanan vicariance and long-distance dispersal. *Journal of Biogeography*, 41: 894-904.

Wielgorskaya, T.V., 2010. Armen Leonovich Taktajan (1910-2009). *Studies in the History of Biology*, 2(2): 8-31 (in Russian).

Vavilovia Federov, *Trudy Biol. Inst. Arm. Fil. Akad. Nauk URSS* 1: 45. 1939 (Fabaceae [Leguminosae], Fabales, 'Fabids')

Golubev, A.A., 1990. Habitats, collection, cultivation and hybridization of *Vavilovia formosa*. *Trudy no Prikladnoy Batanike, Genetike i Selektsii*, 135: 67-75.

Kenicer, G. et al., 2009. *Vavilovia formosa*, an intriguing Pisum relative. *Grain Legumes*, 51: 8.

Mikić, A. et al., 2013. The bicentenary of the research on 'beautiful' vavilovia (*Vavilovia formosa*), a legume crop wild relative with taxonomic and agronomic potential. *Botanical Journal of the Linnean Society*, 172: 524-531.

Pringle, P., 2009. *The Murder of Nikolai Vavilov: the story of Stalin's persecution of one of the twentieth century's greatest scientists*. JR Books, London.

Schaefer, H. et al., 2012. Systematics, biogeography, and character evolution of the legume tribe Fabeae with special focus on the middle-Atlantic lineages. *BMC Evolutionary Biology*, 12: 250

Vickia Roque & Sancho, *Taxon* 69(4): 6701. 2020 (Asteraceae [Compositae], Asterales, 'Campanulids')

Funk, V.A. and Brooks, D.R. (eds.), 1981. *Advances in cladistics: proceedings of the first meeting of the*

Willi Hennig Society. New York Botanical Garden, Bronx.

Funk, V.A. et al., 2014. A phylogeny of the Gochnatieae: understanding a critically placed tribe in the Compositae. *Taxon*, 63(4): 859-882.

Funk, V. et al., (eds.), 2009. *Systematics, Evolution and Biogeography of the Compositae*. International Association for Plant Taxonomy, Vienna.

Gillespie, R.G. and Whittaker, R.J., 2020. Vicki Ann Funk (1947-2019): a short tribute. *Frontiers in Biogeography*, 12: 1-2.

Hennig, W., 1966. *Phylogenetic Systematics* [translation from the German *Grundzüge einer Theorie der phylogenetischen Systematik*, 1950]. University of Illinois Press, Champaign.

Kapsalis, E., 2019. *Because of Her Story: the Funk List*. Online at https://womenshistory.si.edu/ news/2019/11/because-her-story-funk-list

Roque, N. and Funk, V.A., 2013. Morphological characters add support for some members of the basal grade of Asteraceae. *Botanical Journal of the Linnean Society*, 171: 568-586.

Roque, N. and Sancho, G., 2020. *Vickia*, a new genus of tribe Gochnatiae (Compositae). *Taxon*, 69(4): 668-178.

Wagner, W.L. and Specht, C.D., 2020. Vicki Ann Funk (1947-2019), influential Smithsonian botanist. *Taxon*, 69(4): 649-654.

Victoria Lindley, *Victoria Regia* 3. 1837 (Nymphaeaceae, Nympheales, 'Basal angiosperms')

D'Orbigny, A., 1840. Note sur les espèces du genre *Victoria*. *Annales de Sciences Naturelles*, ser. 2, 13: 53-57.

D'Orbigny, A., 1835. *Voyage dans l'Amerique meridionale [...] exécuté pendant les années 1826, 1827, 1828, 1829, 1830, 1832, et 1833*. Tome 1. Chez Pitois-Levrault & Cie., Paris.

Gray, J.E., 1839. [Meeting of] September 7th. *Proceedings of the Botanical Society of London* 1: 44-47.

Guillemin, J.B.A., 1840. Observations sur les genres Euryale et Victoria. *Annales de Sciences Naturelles*, ser. 2, 13: 50-52.

Hooker, W.J., 1847. *Description of Victoria Regia, or Great water-lily of South America*. Reeve Brothers, London.

Klotzsch, J.F., 1847. Falsche Namen in Zeitungen. *Botanische Zeitung*, 5(14): 244-245.

Les, D.H. et al., 1999. Phylogeny, classification, and floral evolution of water lilies (Nymphaeaceae: Nymphaeales): a synthesis of non-molecular, *rbcL*, *matK*, and 18S rDNA data sets. *Systematic Botany*, 24(1): 28-46.

Lindley, J., 1837. *Victoria Regia*. Privately published, hand-coloured folio.

Nielsen, D., 2010. Victoria regia's bequest to modern architecture. In: *Design and Nature V. Comparing design in nature with science and engineering*. WIT Transactions on Ecology and the Environment, 28-30 June 2010.

Poeppig, E.F., 1832. Doctor Poeppig's naturhistorische Reisenberichte*). *Notizen aus dem Gebiete der Natur-und Heilkunde*, 15(9): 129-135.

Prance, G.T., 1974. *Victoria amazonica* ou *Victoria regia*? *Acta Amazonica*, 4(3): 1-8.

Prance, G.T. and Arias, J.R., 1975. A study of the flora biology of *Victoria amazonica* (Poepp.) Sowerby (Nymphaeaceae). *Acta Amazonica*, 5(2): 109-139.

Schomburgk, R.H., 1837. Diary of an ascent of

the River Berbice, in British Guayana, in 1836-7. *Journal of the Royal Geographical Society of London*, 7: 302-350.

Sowerby, J de C., 1850. On the names of the Victoria water lily. *Annals and Magazine of Natural History*, ser. 2, 6: 310.

Thien, L.B. et al., 2009. Pollination biology of basal angiosperms (ANITA grade). *American Journal of Botany*, 96: 166-182.

Washingtonia H.Wendland, *Bot. Zeitung (Berlin)* 37: 68. 1879, nom. cons. (Arecaceae [Palmae], Arecales, 'Commeliniids')

Parish, S.B., 1907. A contribution toward a knowledge of the genus *Washingtonia*. *Botanical Gazette*, 44: 408-434.

Rehder, A. et al., 1935. Conservation of later generic homonyms. *Bulletin of Miscellaneous Information (Royal Botanic Gardens, Kew)*, 1935: 341-544.

Wendland, H., 1879. Ueber Brahea oder Pritchardia filifera hort. *Botanische Zeitung*, 5: 65-69.

Vogl, R.J. and McHargue, L.T., 1966. Vegetation of California fan palm oases on the San Andreas fault. *Ecology*, 47:532-540.

Wuacanthus Y.F. Deng, N.H. Xia & H. Peng, *Plant Diversity* 38(6): 318. 2016 (Acanthaceae, Lamiales, 'Lamiids')

Boufford, D.E., 2014. Biodiversity hotspot: China's Hengduan Mountains. *Arnoldia*, 72: 24-35.

Deng, Y-F. et al., 2016. *Wuacanthus* (Acanthaceae), a new Chinese endemic genus segregated from *Justicia* (Acanthaceae). *Plant Diversity*, 38: 312-321.

Hong, D-Y. and Blackmore, S., 2013. *Plants of China: a companion to the Flora of China*. Science Press, Beijing.

Liang, Q-L. et al., 2018. Shifts in plant distributions in response to climate working in a biodiversity hotspot, the Hengduan Mountains. *Journal of Biogeography*, 45: 1334-1344.

Lv, C-C., 2016. Chronicle of Wu Zhengyi. *Plant Diversity*, 38: 330-344.

Raven, P.H., 2016. Appreciation of a great man: Wu Zhengyi (1916-2013). *Plant Diversity*, 38: 262-263.

Sun, H. et al., 2016. Origins and evolution of diversity in the Hengduan mountains, China. *Plant Diversity*, 38: 161-166.

Wu, Z-Y., 1988. Hengduan mountain flora and her significance. *Journal of Japanese Botany*, 63: 297-311.

Xing, Y. and Ree, R.H., 2017. Uplift-driven diversification in the Hengduan Mountains, a temperate biodiversity hotspot. *Proceedings of the National Academy of Sciences, USA*, 114: e3444-e3451.

Ye, X. et al. 2015. Assessing local and surrounding threats to the protected area network in a biodiversity hotspot: the Hengduan Mountains of southwest China. *PLoS ONE*, 10(9): e0138533

Yu, H. et al., 2020. Contrasting floristic diversity of the Hengduan Mountains, the Himalayas and the Qinghai-Tibet Plateau sensu stricto in China. *Frontiers in Ecology and Evolution*, 6: 136.

Zhou, Z. and Sun, H., 2016. Wu Zhengyi and his contributions to plant taxonomy and phytogeography. *Plant Diversity*, 38: 259-261.

Acknowledgements

I have been thinking about a book like this – linking plants with the people for whom they are named – for a very long time, but the various lockdowns due to the Covid-19 pandemic put a stop to my fieldwork and travel schedules giving me the space to really think about it. So many people have shared their knowledge and insights about both people and plants over the course of my writing this book. Others have listened patiently to my tales of daring and wonder, others have accompanied me in discovering for myself some of the places where these plants grow, others have talked to me about having plants named for them, and still others have patiently read drafts at various stages and/ or contributed images of plants and people. I apologise in advance to anyone I have inadvertently omitted. Thank you all so much for all your help: David Attenborough, Robbie Blackhall-Miles, Will Beharrell, Mauricio Bonifacino, Isabelle Charmantier, Seth Cotterell, Helena Crouch, Thomas Couvreur, Chuck Davis, Deng Yun-Fei, Soejatmi Dransfield, Socorro González-Elizondo, Andrea Hart, Nick Helme, Peter Hoch, Doug Holland, Kanchi Gandhi, Robert Knapp, Fay Wei Li, H. Peter Linder, Aimee McArdle, Jean Yves Hiro Meyer, Fabian Michelangeli, Mary Knapp Parlange, Kathleen Pryer, Peter and Pat Raven, Nadia Roque, Fred Rumsey, George Schatz, Dmitry Telnov, Tong Yi, Maria (Bat) Vorontsova, Warren Wagner, Jacek Wajer, Mark Watson, Jun Wen – without all of you this book would not be what it is. That said, all mistakes are my own! At the Natural History Museum, the Publishing team – Trudy Brannan, Lynn Millhouse, Lewis Morgan, Gemma Simmons, Anna Smith and Colin Ziegler – have all put with my probably rather irritating demands for this or that, at Chicago, Joseph Calamia has let me think aloud on walks around London parks during various lockdowns. The California Academy of Sciences, the Linnean Society of London and the Royal Botanic Gardens of Edinburgh and Kew have kindly allowed me use images under their copyright (credits for all images are below). My twin loves of plants and literature were always encouraged by my late parents Ed and Jean Knapp, their encouragement for me to follow my dreams has allowed me to embrace the study of plants with open heart and mind. My special thanks go to my family, Victor, Isabel, Alfred, Charlotte, Libby and Julia – you are the generations in whose hands the future of plants rests.

Picture credits